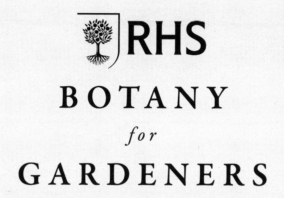

BOTANY
for
GARDENERS

RHS Botany for Gardeners
Contributing Author: Geoff Hodge
RHS Consultant Editor: Simon Maughan

First published in Great Britain in 2013 by Mitchell Beazley,
an imprint of Octopus Publishing Group Ltd,
Carmelite House, 50 Victoria Embankment, London EC4Y 0DZ
www.octopusbooks.co.uk

An Hachette UK Company
www.hachette.co.uk

Published in association with the Royal Horticultural Society

2013 © Quarto Publishing plc

All rights reserved. No part of this work may be reproduced or utilized in any form or by any means, electronic or mechanical, including photocopying, recording or by any information storage and retrieval system, without the prior written permission of the publishers.

ISBN: 978 1 84533 833 6

12

A CIP record of this book is available from the British Library

Set in Garamond Pro

Printed and bound in China

Mitchell Beazley Publisher: Alison Starling
RHS Publisher: Rae Spencer Jones

Conceived, designed and produced by
The Bright Press, an imprint of The Quarto Group.
1 Triptych Place, London
SE1 9SH
(0)20 7700 6700   www.quarto.com

Design: Lindsey Johns
Technical illustration: Sarah Skeate

The Royal Horticultural Society is the UK's leading gardening charity dedicated to advancing horticulture and promoting good gardening. Its charitable work includes providing expert advice and information, training the next generation of gardeners, creating hands-on opportunities for children to grow plants and conducting research into plants, pests and environmental issues affecting gardeners.
For more information visit *www.rhs.org.uk* or call 0845 130 4646.

RHS
BOTANY
for
GARDENERS

The Art and Science of Gardening
Explained and Explored

MITCHELL
BEAZLEY

Contents

How to use this book	6
A short history of botany	8

CHAPTER 1
THE PLANT KINGDOM

Algae	12
Mosses and liverworts	14
Lichens	18
Ferns and their relatives	19
Gymnosperms: conifers and their relatives	22
Angiosperms: flowering plants	25
Monocotyledons versus dicotyledons	28
Plant naming and common names	29
Plant families	31
Genus	34
Species	36
Hybrids and cultivars	39

CHAPTER 2
GROWTH, FORM AND FUNCTION

Plant growth and development	44
Buds	51
Roots	56
Stems	62
Leaves	66
Flowers	71
Seeds	74
Fruit	78
Bulbs and other underground food storage organs	82

CHAPTER 3
INNER WORKINGS

Cells and cell division	86
Photosynthesis	89
Plant nutrition	91
Nutrient and water distribution	96
Plant hormones	98

CHAPTER 4
REPRODUCTION

Vegetative reproduction	102
Sexual reproduction	110
Plant breeding – evolution in cultivation	118

CHAPTER 5
THE BEGINNING OF LIFE

Development of the seed and fruit	124
Seed dormancy	125
Seed germination	126
Sowing and saving seeds	132
Seed saving	134

Dahlia × hortensis, dahlia

CHAPTER 6
EXTERNAL FACTORS

The soil	138
Soil pH	144
Soil fertility	145
Soil moisture and rainwater	148
Nutrients and feeding	152
Life above ground	153

CHAPTER 7
PRUNING

Why prune?	160
Pruning trees	164
Pruning for size and shape	170
Pruning for display	172

CHAPTER 8
BOTANY AND THE SENSES

Seeing light	178
Sensing scent	184
Scent as an attractant	185
Feeling vibrations	186

CHAPTER 9
PESTS, DISEASES AND DISORDERS

Insect pests	190
Other common pests	194
Fungi and fungal diseases	198
Viral diseases	203
Bacterial diseases	205
Parasitic plants	207
How plants defend themselves	209
Breeding for resistance to pests and diseases	214
Physiological disorders	215

Aloe brevifolia, short-leaved aloe

Index	220
Bibliography	223
Credits and websites	224

BOTANISTS AND BOTANICAL ILLUSTRATORS

Gregor Johann Mendel	16
Barbara McClintock	32
Robert Fortune	54
Prospero Alpini	60
Richard Spruce	76
Charles Sprague Sargent	94
Luther Burbank	108
Franz and Ferdinand Bauer	116
Matilda Smith	130
John Lindley FRS	150
Marianne North	168
Pierre-Joseph Redouté	182
James Sowerby	196
Vera Scarth-Johnson OAM	218

How to use this book

Botany for Gardeners is written for those interested in gardening but with a desire to dip a toe into the science behind plants. The science is kept at a level so as not to be indecipherable, and any botanical language used is always qualified with an explanation. Furthermore, the authors have been careful never to stray too far from the interests of the practical gardener, and as a result many examples used to illustrate the text come from plants that a gardener might know, and might even have grown themselves. 'Botany in Action' text boxes can be found throughout the book and highlight information that is of particular practical interest to gardeners.

The book is structured into nine chapters, each dealing with an important area of botany as relevant to gardeners. Consequently, there are chapters on the plant kingdom and plant naming (chapter 1), seed germination and growing (chapter 5), as well as a botanical look at pruning (chapter 7). Chapters 6 and 9 move beyond the realm of botany into the very closely related subjects of soil science, plant pathology and entomology. *Botany for Gardeners* is not designed to be read in any particular order, rather each chapter is almost a subject in its own right, and where information touches on that of another chapter, clear cross references are supplied.

Periodically through the book, the lifetime achievements of various botanists and botanical illustrators are introduced. This serves to remind the reader of the historical context of botany, and the debt gardeners owe to the pursuits of plant scientists over the centuries. The fifteen botanists chosen are not intended to be a definitive list by any means – the history of botany is populated by a huge number of fascinating characters who made equally fascinating discoveries and sometimes struggled to have their ideas accepted. It is a subject that deserves further study.

While *Botany for Gardeners* is intended to inform gardeners, the practical examples and practical advice given are not intended to be comprehensive. Readers will find many pests and diseases discussed in chapter 9, and some suggested treatments; likewise, in chapter 7, many pruning cuts are described. Gardeners wishing to explore the practical side of these subjects in more detail are advised to read further afield; overall it is hoped that this book will inspire a greater understanding of the subject and a lifetime of informed gardening.

Prunus persica,
peach

Prunus is a large genus of ornamental and edible plants, including cherry and plum. The species name persica refers to Persia (now Iran) from where it came to Europe.

HOW TO USE THIS BOOK

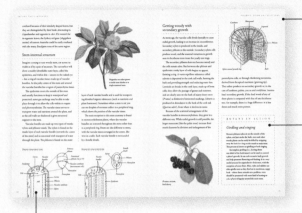

MAIN PAGES
All nine chapters are led by the content of these core pages. Clear introductions and category headings make the text easy to understand, whilst accompanying plant illustrations are annotated with both Latin and common names.

BOTANY IN ACTION
Short feature boxes placed throughout the book demonstrate ways in which the theory can be turned into practice, providing gardeners with practical tips.

DIAGRAMS
As well as dozens of attractive botanical illustrations and etchings, the book includes numerous simple annotated diagrams to clarify technical aspects.

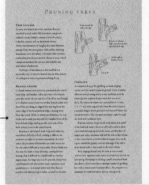

FEATURE PAGES
Dotted throughout the book are feature pages providing a range of practical applications for the gardener in a bitesize form. Examples include Pruning and Breaking Seed Dormancy.

BOTANISTS AND BOTANICAL ILLUSTRATORS
Feature spreads profile notable men and women in the history of botany, exploring their lives and explaining the ways in which their work was influential.

Cycas siamensis,
silver cycad,
Thai silver cycad

Botany ORIGIN late 17th century from earlier *botanic*, from French *botanique*, based on Greek *botanikos*, from *botanē*, 'plant'.

The scientific study of plants, including their physiology, structure, genetics, ecology, distribution, classification and economic importance.

A Short History of Botany

The first simple studies of plants started with early man, the hunter-gatherers of the Palaeolithic era, who were the first people to put down roots and start to practise agriculture. These studies were basic interactions, with plant lore passed down from generation to generation that identified which plants were nutritious and could be eaten, and which were toxic. Further interactions included using plants as herbal remedies and cures for illnesses and other problems.

The first physical records about plants were made around 10,000 years ago, as the written word developed as a means of communication, but the true study of plants started with Theophrastus (371–286 BC), who is known as the father of botany. He was a student of Aristotle, and is considered the starting point for research into plants and hence botany. He wrote many books, including two important sets of books on plants – *Historia de Plantis* (*History of Plants*) and *De Causis Plantarum* (*On the Causes of Plants*).

Theophrastus understood the difference between dicotyledons and monocotyledons, and angiosperms and gymnosperms (see pp .22–29). He categorised plants into four groups: trees, shrubs, undershrubs and herbs. He also wrote on important subjects such as germination, cultivation and propagation.

Pedanius Dioscorides was another important name in early botany. He was a physician and botanist in Emperor Nero's army. He wrote the five-volume encyclopedia *De Materia Medica* (*Regarding Medical Materials*) between 50 and 70 AD, which dealt with the pharmacological uses of plants. This was the most influential work until the 1600s, and served as an important reference work for later botanists.

In medieval Europe, botanical science took a back seat and was overshadowed by the preoccupation with the medicinal properties of plants, and herbals became the standard works in which plants were studied and written about, probably the best-known being Culpeper's *Complete Herbal & English Physician*.

It was not until the European Renaissance between the 14th and 17th centuries that botany experienced a resurrection and resumed its importance in the study of nature and the natural world, emerging as a science in its own right. The herbals were supplemented by floras, consisting of more detailed accounts of the native plants of a region or country. The invention of the microscope, in the 1590s, encouraged the detailed study of plant anatomy and sexual reproduction, and the first experiments in plant physiology.

Lonicera × *brownii*,
Brown's honeysuckle, scarlet trumpet honeysuckle

A semi-evergreen, climbing honeysuckle – a hybrid from a cross between *Lonicera sempervirens* and *L. hirsuta*.

As world exploration and trading with countries farther afield became more widespread, many new plants were discovered. These were often cultivated in European gardens, some becoming new food staples, and their accurate naming and classification became very important.

In 1753, only a little more than a century before Darwin published *The Origin of Species*, Carl Linnaeus published his *Species Plantarum*, one of the most important works in biology. Linnaeus' work contained the known plant species of the time. He created a system for organising plants in a uniform manner, so that anyone could find and name a plant based on its physical characteristics. He grouped plants and gave every plant a binomial (two-part) name, and so began the universal binomial nomenclature system still in use today.

A gathering number of scientists gradually began to contribute to Linnaeus' work, leading to an enormous growth in plant knowledge, as more and more discoveries were made. The scientists who were making these discoveries became increasingly specialised, leading to yet further discoveries.

During the 19th and 20th centuries, the use of more sophisticated scientific technology and methods expanded the knowledge of plants exponentially. The 19th century set the foundations of modern botany. Research was published in papers by research schools, universities and institutes (rather than just being the domain of an elite few 'gentlemen scientists'), so that all this new information was available to a much wider audience.

In 1847, the theory concerning the role of photosynthesis in capturing the sun's radiant energy was first discussed. In 1903, chlorophyll was separated from plant extracts, and during the period from the 1940s to the 1960s, the complete mechanism of photosynthesis became fully understood. New areas of study began, including the practical fields of economic botany – agriculture, horticulture and

Alyogyne hakeifolia is found in southern areas of Australia. The genus *Alyogyne* is similar to *Hibiscus*.

forestry – as well as extremely detailed studies of the structure and function of plants, such as biochemistry, molecular biology and cell theory.

By the 20th century, radioactive isotopes, electron microscopes and a wealth of new technologies, including computers, all helped in the understanding of how plants grow and react to changes in their environment. At the close of the millennium, the genetic manipulation of plants was a hot topic of discussion and this technology is likely to play a major role in the future of the human race.

It is clear, though, from writing and researching this book, that there is still a considerable amount about plants that we do not yet know. It is sobering to realise that the mysteries of photosynthesis have only just been revealed in the last 60 years. Waiting out there, among the hundreds of thousands of plant species, are many, many more secrets yet to be revealed.

Lilium pensylvanicum,
Siberian lily

Chapter 1
The Plant Kingdom

For the study of nature to be possible, humans have long sought to arrange the great diversity of living things into groups that bear similar characteristics. This is known as classification, and depending on the system used, all living things are split into a number of major groups known as kingdoms.

From a gardener's perspective, the starting point for plant classification begins with the question, 'Is it a tree, a shrub, a perennial, or a bulb?' Botanists also recognise these groups, although they are not used as a basis for taxonomy (scientific classification) – in other words, the plant kingdom is not classified scientifically along these lines.

The organisms within the plant kingdom are classified according to their evolutionary groups, starting with the more simple algae and ending with the more highly developed flowering plants. With a few exceptions, all organisms within this kingdom share the ability to manufacture their own food from sunlight, through photosynthesis.

At a first glance plant classification can seem confusing. However, knowing how plants are classified will help you achieve a greater appreciation of what you grow in your garden and provide a sound basis for further study. In this chapter, the main groupings into which the plant kingdom has been separated are discussed.

Algae

It has to be said that gardeners may have very limited interest in algae (singular: alga). Other than pond algae, and the slippery slime that can accumulate on damp decks and patios, these organisms play a minor role in gardeners' minds.

Before we dismiss algae, however, it is worth mentioning that much of the plant kingdom is made up of these simple life forms, and they play a massively important role in the world's ecosystem. They are considered 'simple' because they lack the many different cell types of other plants and do not possess complex structures such as roots, leaves and other specialised organs.

Diatoms are common algae; they occur almost everywhere that is adequately lit and moist – ponds, bogs and damp moss. They are among the most common phytoplankton and most are single-celled.

Huge variety is seen within this group of organisms. Most of us will be familiar with seaweeds, which are multicellular algae, but also prevalent are single-celled phytoplankton, which fill the seas and generate food using the sun's energy, thereby supporting all marine life. One curious group of algae are the diatoms: microscopic, single-celled algae that are an ever-present, yet invisible feature of any watery habitat. They are encased in fascinatingly beautiful, silicon-based cell walls.

As would be expected from such a 'simple' form of life, the reproductive strategies of algae are not as complex as those seen in higher plants. Mostly, algae reproduce vegetatively, through the splitting of individual cells or larger multicellular units, and sexual reproduction is achieved by the meeting and ultimate fusion of two mobile cells.

Ascophyllum nodosum,
knotted kelp or knotted wrack

This common brown seaweed, also known as Norwegian kelp, is used to make plant fertilisers and in the manufacture of seaweed meal.

Typical algae reproduction

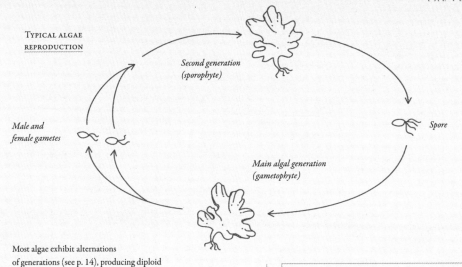

Most algae exhibit alternations of generations (see p. 14), producing diploid sporophytes and haploid gametophytes.

Algae in the garden

Because algal cells do not produce a waterproof cuticle or have other means to prevent themselves drying out, they are either found in water or in damp, shady places. They also need the constant presence of water for growth and reproduction.

In the garden, algae will almost certainly be found in any garden pond or other area of standing water or constant moisture. Algae will also be found in the soil.

Pond algae

The pond is where most gardeners will encounter algae, and they can become quite a problem, especially when the weather warms up in spring. If conditions are favourable, algae can quickly discolour pond water, form unsightly scums or choke the water with filamentous growth (blanket weed). Left to their own devices, algae can deprive the water of oxygen to the detriment of other pond life.

Despite this, algae are an essential part of the natural food chain in water gardens, and when kept 'in balance', they help to maintain a healthy water environment. Problems seem to occur when ponds are exposed to too much sun, when temperatures fluctuate too widely (particularly problematic in small ponds), and where nutrient levels are too high. High nutrient levels may be caused by a build-up of debris in the pond and on the pond floor, as well as by fertilisers leaching into the water.

Hard-surface algae

Algae will also grow on wet paths, fences, garden furniture and other hard surfaces, especially those in cool, shady areas. Mosses, lichens and liverworts may also be present in such situations. Contrary to popular belief, algae do not damage the hard surfaces on which they grow (though they may leave stains or marks), but they can make surfaces very slippery and treacherous. It is therefore worth trying to remove them, either with a pressure washer or a proprietary path and patio cleaner.

BOTANY IN ACTION

Removing pond algae

It is very difficult to totally eliminate pond algae. Chemical controls will cause algae to die, but they will then rot and make the problem worse as more nutrients build up in the water. A better option is to install a filter to remove the nutrients and the algae.

Mosses and liverworts

To botanists, this group of plants are known as the bryophytes. They are generally restricted to moist habitats; many are in fact aquatic. As multicellular organisms, these plants are considered more advanced than algae. However, they are still relatively simple plants with little differentiation between the cells, though some do have specialised tissues for the transport of water.

To the gardener, mosses are more significant than liverworts, as they are commonly seen in almost all gardens, tending to grow in wet or damp, shady places in clumps or mats. Sphagnum moss is a major component of peat, still used in plant potting compost, although gardeners are now encouraged to buy peat-free alternatives. Liverworts are less noticed by the gardener; they are quite different to mosses in appearance, having a flattened, leathery body, which is sometimes lobed. Mosses have a more elaborate structure than liverworts, often with upright shoots bearing tiny 'leaflets'. In common with algae, bryophytes can only reproduce sexually in the presence of water. Without the medium of water, the male and female sex cells (sperm and egg) would not be able to meet.

The alternation of generations

With bryophytes, we see the appearance of a complex life cycle, known as 'the alternation of generations', which is a phenomenon seen in all plants beyond a certain level of complexity. There are two generations to the life cycle: gametophyte and sporophyte. In mosses and liverworts, the plant spends the majority of its life cycle in the gametophyte stage; in ferns and all higher plants the sporophyte stage is dominant. In flowering plants, the gametophyte stage is so reduced that it is often not referred to in these terms (see p. 22).

In the gametophyte stage, each and every cell carries just half of the organism's genetic material. Thus, the structures we know as mosses or liverworts

Bryophytes have a multicellular and structured main body. They produce enclosed reproductive structures and spread via spores produced within a structure called the sporangium.

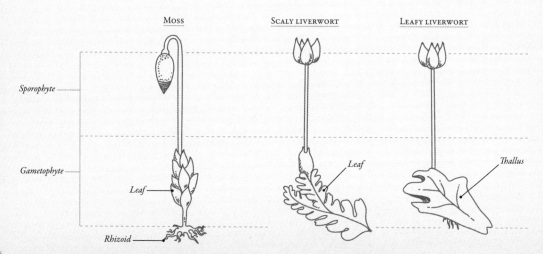

THE PLANT KINGDOM

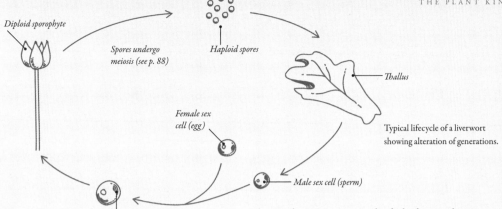

Typical lifecycle of a liverwort showing alteration of generations.

are actually just made up of unpaired 'half cells' (haploids). It is only when these structures release sperm and egg cells, which meet and fuse in the presence of water, that a 'whole cell' (diploid) is brought into being. This becomes the sporophyte generation, and in bryophytes the sporophyte generation is reduced to a simple spore-producing body that remains attached to the gametophyte.

Bazzania trilobata, greater whipwort, is a moss. Both generations are visible here.

As the name suggests, the diploid sporophyte generation releases spores, which are created by the division of sporophyte cells. Thus, the spores themselves are haploid, and upon release they are dispersed by the rain or wind, after which some of them will grow into a new moss or liverwort gametophyte.

BOTANY IN ACTION

Mosses and liverworts in the garden

Although mosses are more often seen as a problem by the gardener, such as when they infest lawns, block guttering and cause unsightly growth on paving and wooden structures, some do have ornamental uses. In Japanese-style gardens, mosses are used to adorn old structures, and moss is also widely used in bonsai as a soil covering as well in hanging baskets as a moisture-holding retaining material. The trend for green roofs is also extending its use. However, mosses can be extremely difficult to maintain and cultivate away from their natural habitats, as they often have very particular requirements regarding light, humidity and substrate chemistry.

Brick, wood and concrete surfaces, including hypertufa, all make potentially good surfaces for moss; they can be prepared, to make them more hospitable, using substances like milk, yoghurt, manure, or a mixture of all three.

Liverworts can become a problem on soil in shady areas or in pots. Where they cannot be tolerated, or better still ignored, they are treated as weeds.

Gregor Johann Mendel
1822–1884

Gregor Mendel is famous for his experiments on the inheritance of physical traits in plants.

Gregor Johann Mendel, now regarded as the father and founder of the science of genetics, was born Johann Mendel in what was then Heinzendorf, Austria, which is now in the Czech Republic.

He lived and worked on the family farm and during his childhood mainly spent time in the garden and studying beekeeping. He went on to attend the Philosophical Institute of the University of Olmütz, where he studied physics, maths and practical and theoretical philosophy and distinguished himself academically. The head of the university's Natural History and Agriculture Department was Johann Karl Nestler, who was conducting research into the hereditary traits of plants and animals.

During his graduation year, Mendel began studying to be a monk, and joined the Augustinian order at the St Thomas Monastery in Brno where he was given the name Gregor. This monastery was a cultural centre and Mendel soon became involved in the research and teaching of its members and had access to the monastery's extensive library and experimental facilities.

After eight years at the monastery Mendel was sent to the University of Vienna, at the monastery's expense, to continue his scientific studies. Here he studied botany under Franz Unger, who was using microscopes and was a proponent of a pre-Darwinian version of evolutionary theory.

After completing his studies at Vienna, Mendel returned to the monastery where he was given a teaching position at a secondary school. During this time he began the experiments that would make his name famous.

Mendel began to research the transmission of hereditary traits in plant hybrids. At the time of Mendel's studies, it was generally accepted that the hereditary traits of the offspring were simply the diluted blending of whatever traits were present in the parents. It was also commonly accepted that, over several generations, a hybrid would revert to its original form, suggesting that a hybrid could not create new forms. However, the results of such studies were usually skewed by short experimental times. Mendel's research continued for up to eight years and involved tens of thousands of individual plants.

Mendel used peas for his experiments because they exhibit many distinct characteristics, and because offspring could be quickly and easily produced. He cross-fertilised pea plants that had clearly opposite

'MY SCIENTIFIC STUDIES HAVE AFFORDED ME GREAT GRATIFICATION; AND I AM CONVINCED THAT IT WILL NOT BE LONG BEFORE THE WHOLE WORLD ACKNOWLEDGES THE RESULTS OF MY WORK.'
Gregor Mendel

characteristics, including tall with short, smooth seeds with wrinkled seeds, green seeds with yellow seeds. After analysing his results, he showed that one in four pea plants had purebred dominant genes, one had purebred recessive genes and the other two were intermediates.

These results led him to two of his most important conclusions and the creation of what would become known as Mendel's Laws of Inheritance. The Law of Segregation reasoned that there are dominant and recessive traits passed on randomly from parents to offspring. The Law of Independent Assortment established that these traits were passed on independently of other traits from parent to offspring. He also proposed that this heredity followed basic mathematical statistical laws. Although Mendel's experiments involved peas, he put forward the hypothesis that this was true for all living things.

In 1865, Mendel delivered two lectures on his findings to the Natural Science Society in Brno, who published the results of his studies in its journal under the title 'Experiments on Plant Hybrids'. Mendel did little to promote his work and the few references to his work from that time indicate that much of it had been misunderstood. It was generally thought that Mendel had only demonstrated what was already commonly known at the time – that hybrids eventually revert to their original form. The importance of variability and its implications were overlooked.

In 1868, Mendel was elected abbot of the school where he had been teaching for the previous 14 years, and both his increased administrative duties and failing eyesight stopped him carrying out further scientific work. His work was still largely unknown and somewhat discredited when he died.

It was not until the early 1900s, when plant breeding, genetics and heredity became an important area of research, that the significance of Mendel's findings became fully appreciated and recognised and began to be referred to as Mendel's Laws of Inheritance.

Lathyrus odoratus,
sweet pea

Peas were the subjects for Mendel's famous genetics experiments – they show several distinct characteristics, and young plants can be produced quickly and easily.

Lichens

Only 150 years ago did scientists discover the true nature of lichens. They are a curious partnership between fungi and algae, living together in a symbiotic relationship. Today they are classified by their fungal component, which puts them outside of the plant kingdom, but they are included here as they have long been a subject of botanical study.

Lichens seem to be able to grow in every habitat on Earth, and in some extreme environments, such as on exposed rock in polar climates, they seem to be the only thing that can grow. In 2005, scientists even discovered that two species of lichen were able to survive for 15 days exposed to the vacuum of space. More commonly, they are seen growing on trees and shrubs, bare rock, walls, roofs and paving and on the soil. Informally, they are generally divided into

Lichens have variable appearances. Some look like leaves (foliose), others are crust-like (crustose), adopt shrubby forms (fruticose) or are gelatinous.

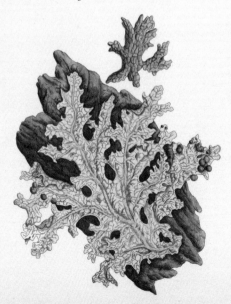

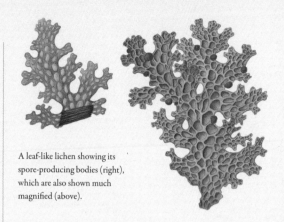

A leaf-like lichen showing its spore-producing bodies (right), which are also shown much magnified (above).

seven groups depending on how they grow; thus we have crustose, filamentous, foliose (leafy), fruticose (branched), leprose (powdery), squamulose (scaly), and gelatinous lichens.

Lichens in the garden

More often than not, lichens are noticed on lawns, where their appearance, quite rightly, often causes gardeners concern. Lichens not only affect the appearance of the lawn, they block light from reaching the grass (so killing it), and they can make surfaces slippery.

In turf, the most common lichens are the dog lichens (*Peltigera*). They are either dark brown, grey or nearly black, and are formed of flat structures that grow horizontally in the turf. They are usually worse on lawns with poor drainage, compacted soil and shady conditions, and as they grow in similar conditions to moss, the two often appear together. Interestingly, dog lichens have the ability to fix atmospheric nitrogen, so they are beneficial in soil fertility.

To prevent lichens on a lawn, you need to improve drainage, thereby correcting the underlying conditions that enabled them to grow in the first place. There are few, if any, effective chemical controls available to gardeners, but path and patio cleaners can be used to scrub them off hard surfaces.

Ferns and their relatives

From an evolutionary point of view, ferns and their relatives represent a significant development: when plants began to show increased cell differentiation. Here we see the first vascular systems – vessels for transporting water and nutrients around the plant – as well as structures concerned with the support of the plant. These were also the first plants to truly colonise the land.

Botanists classify this group of plants as pteridophytes, and they include club mosses, true ferns and horsetails. Gardeners are likely to have heard of horsetails, and they will definitely know about ferns, but club mosses (sometimes called spike mosses) remain obscure, even though one or two forms are cultivated. Club mosses are not mosses; they are more advanced.

Selaginella martensii (little club moss) produces trailing stems and is a good ground-cover plant for moist, shady places.

Like the bryophytes, pteridophytes exhibit a clear alternation of generations, but the crucial shift is that pteridophytes spend most of their life cycle in the sporophyte phase. This allows for the production of vertical branches or fronds, sometimes specialised, bearing tiny swellings called sporangia. When the sporangia burst, they release the spores, which germinate to become the gametophyte generation.

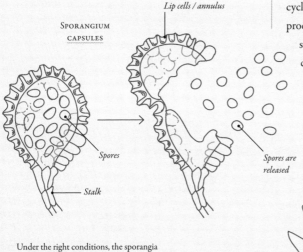

Under the right conditions, the sporangia burst open, releasing the spores. These are carried by the wind and grow into the gamete-producing gametophyte.

Underside of fern frond showing the spore-producing sporangia.

A gardener could be easily forgiven for assuming that spores are simply the same as seeds. While they are both used as ways for a plant to disperse themselves and there are similarities in their culture, it is important to note that there are vital differences: a spore is generally much smaller than a seed and its production does not rely on fertilisation. Ferns do not produce seeds.

Allowed to grow on a tray of seed compost, and given sufficient moisture as well as the required amount of light and heat, fern spores will begin to grow. But rather than growing into baby ferns, they will grow into the next stage of the life cycle: the gametophyte generation. These odd-looking plants are called prothalli, and if kept moist and misted they will slowly begin to grow into new ferns – what the human eye does not see is that during this interval, the prothallus has itself produced sperm cells that have fertilised the egg cells (this is the stage of sexual reproduction), which then grown into the new sporophyte generation of the fern.

Ferns in the garden

There are approximately 10,000 species of fern, all varying quite remarkably in size and growth habit, from the stately royal fern (*Osmunda regalis*) to the tiny, floating aquatic water fern (*Azolla filiculoides*). The water fern is considered invasive in some parts of the world, due to its prolific nature, while in others it is highly valued in agriculture for enhancing the growth rate of crops grown in water, such as rice. Either way, *Azolla* is a highly successful plant, and one that needs to be avoided like the plague in a garden setting. Bracken (*Pteridium aquilinum*) is a land fern with a similarly invasive nature, and is considered to have the widest worldwide distribution of any fern.

There are many, many more fern species that are commonly used as ornamental garden and indoor plants, and from them plant breeders have selected countless cultivars with variations in frond form and colour. Most ferns grow in moist, shady woodlands and these are the conditions in which they tend to grow best in the garden.

In recent years, some of the most popular ferns in gardens have been those referred to as tree ferns. Any fern that grows with the fronds elevated above ground level by a trunk can be called a tree fern, and in cool

***Pteridium aquilinum*,**
bracken

Bracken can easily become invasive on cultivated ground. It contains a carcinogenic compound that can cause death in livestock.

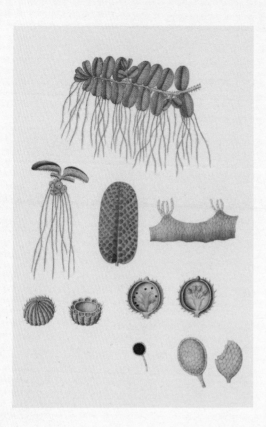

Azolla filiculoides,
water fern

climates perhaps the most familiar example would be *Dicksonia antarctica* from Australia. The 'trunk' is not like that of a tree or shrub; it is in fact a mass of fibrous roots that are built up over time as the crown of the fern continues to grow. In the wild, many tree fern species are threatened with extinction because of deforestation.

Fern relatives

The most notable relatives of the true ferns must be the horsetails (*Equisetum*). Although a handful of species (such as *E. hyemale* and *E. scirpoides*) are grown as ornamental plants, *Equisetum* is best known for *E. arvense*, field horsetail, which is a notorious weed in many areas of the world. It is very difficult to eradicate, and in a garden situation it becomes a persistent and overbearing nuisance.

The most extraordinary fact about *Equisetum*, however, is its status as a 'living fossil'. It is the only genus alive today of its class, which dominated the understorey of the world's forests approximately 400 million years ago. Fossils found in coal deposits show that some *Equisetum* species reached over 30 m (100 ft) tall.

One of the club mosses, *Selaginella*, is regarded as a botanical curiosity: *Selaginella lepidophylla*, a desert plant known as the resurrection plant, is so named because it will curl up into a tight, brown or reddish ball if allowed to dry out, only to uncurl and turn green again when moistened. *S. kraussiana* (Krauss's spike moss) is grown as an ornamental plant in warm climates; it has a number of cultivated forms and is valued for its rapidly extending low growth, which is useful for groundcover in shade.

Equisetum arvense,
field horsetail, mare's tail

Horsetail produces deep searching roots and, once established, can become a serious weed problem and difficult to control.

Young plant

Mature plant

Gymnosperms: conifers and their relatives

These more complex plant life forms belong to a larger group of plants called spermatophytes, which are essentially all those plants that produce seeds. They all possess complex vascular systems and specialised anatomical structures, such as lignified (woody) tissue for support and cones for reproduction. Spermatophytes include all conifers and cycads (collectively known as gymnosperms), as well as the flowering plants (angiosperms) – which are discussed in the next section (see p. 25).

Ginkgo biloba, maidenhair tree

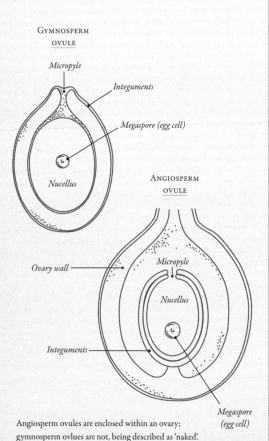

Angiosperm ovules are enclosed within an ovary; gymnosperm ovules are not, being described as 'naked'.

Seeds represent an important development in the evolution of plants, since a major difficulty faced by lower plant forms, such as ferns, is the vulnerability of the delicate gametophyte generation to the external environment. Spermatophytes overcome this problem by protecting the gametophyte within specialised tissue: the female reproductive cells are protected within an ovule, and the male sperm cells are encased within the pollen grain. When the two meet, fertilisation takes place, and the ovule develops into the seed.

The term *gymnosperm* means 'naked seed'. This distinction refers to the inclusion in flowering plants (angiosperms) of the ovules within an ovary; in gymnosperms they are not so enclosed.

Gardeners will know almost any common gymnosperm purely by its physical appearance: these are the conifers and the cycads. The maidenhair tree (*Ginkgo biloba*) is perhaps one exception – with its broad, deciduous leaves it looks most unlike a conifer.

Conifers

Well known for their cones, conifers actually produce two types: small male cones that produce masses of wind-borne pollen, and larger female cones that carry the ovules and ultimately the seeds. The seeds are dispersed in a variety of ways, commonly on the wind or by animals.

A number of unusual variations are seen among particular conifers. Yews (*Taxus*), junipers (*Juniperus*) and plum yews (*Cephalotaxus*) are three examples of 'berry-bearing' conifers. In each case, the seed cones are highly modified, sometimes containing just a single seed surrounded by fleshy arils or modified scales, which develop into a soft, berry-like structure. These attract birds or other animals, which eat them and so distribute the seed.

Compared to flowering plants (angiosperms), the total number of conifer species is low, yet they still manage to dominate swathes of the Earth's surface. They are most abundant throughout the

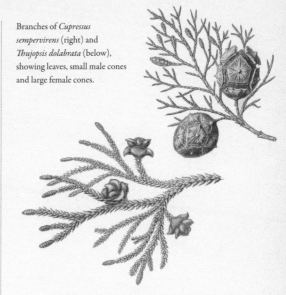

Branches of *Cupressus sempervirens* (right) and *Thujopsis dolabrata* (below), showing leaves, small male cones and large female cones.

Juniperus communis, common juniper

vast boreal forests of the northern hemisphere, with other conifer forests extending into southern regions, particularly in cooler, high-altitude areas. With their high economic value, mainly for their softwood timber used in construction and in paper production, conifers are widely planted in man-made forests throughout the world.

Conifers show a myriad of forms, as a result of evolution as well as artificial breeding. Conifers from cold climates typically exhibit a narrow, conical shape, which helps them to shed snow, while those from high-sunlight areas may have a bluish or silvery tinge to the foliage to reflect ultraviolet light. Plant breeders often exploit mutations or hybridise species to create new forms, often for garden use. Thus we have Leyland cypress (× *Cuprocyparis leylandii*), a hybrid that first occurred in Northern Ireland in about 1870 when a seedling resulted from a cross between Monterey cypress (*Cupressus macrocarpa*) and Nootka cypress (*Xanthocyparis nootkatensis*).

Although most conifers are evergreens, the following are deciduous: larch (*Larix*), false larch (*Pseudolarix*), swamp cypress (*Taxodium*), dawn redwood (*Metasequoia*) and Chinese swamp cypress (*Glyptostrobus*).

Cycads and ginkgo

From a distance, cycads look a lot like palm trees, but on close examination they show clear differences. Their stout, woody trunks are quite different to the fibrous trunks of palms, and their crowns are much more stiffly evergreen. They also bear cones, which is an important difference.

Many, but not all cycads grow very slowly, and specimens more than 2–3 m (6–10 ft) tall are rarely seen in cultivation. They are also extremely long-lived, with some plants known to be at least 1,000 years old. Cycads are considered living fossils, and remain little changed since the Jurassic period, the time of the dinosaurs. Most cycads also have very specialised pollinators, usually a particular species of beetle.

Naturally, cycads grow across much of the subtropical and tropical parts of the world, sometimes from semi-arid to wet rainforests. In cultivation they are only seen in warm-temperate or tropical climates, unless grown under heated glass. Many cycad species

Cycas rumphii, commonly known as the queen sago palm, produces a starchy pith from which sago can be made.

are endangered, due to over-collecting in the wild and destruction of their natural habitat. Some species, such as *Encephalartos woodii*, now exist only in cultivation.

Ginkgos are represented by a single living species, *G. biloba*, the maidenhair tree. It is another living fossil, first appearing in the fossil record some 270 million years ago, yet it is a curiosity, and botanists are unsure of its place among other plants as it has no close relatives. For now it sits among the gymnosperms, as its seeds are not enclosed within an ovary, but the morphology of its 'fruits' confuses the issue. The ginkgo's deciduous leaves are an attractive fan shape and turn a brilliant, butter yellow in autumn as they fall.

Conifers in the garden

While fashions for certain conifers come and go, these plants are a garden mainstay, with their distinctive forms and foliage. Popular genera include the firs (*Abies*), monkey puzzles (*Araucaria*), cedars (*Cedrus*), cypresses (*Cupressus*), junipers (*Juniperus*), pines (*Pinus*), spruces (*Picea*) and false cypresses (*Chamaecyparis*), to name but a few. Literally thousands of hybrids and cultivars of these also exist in cultivation, ranging from dwarf or low-growing shrubs and ground-cover plants to very tall trees, some exceeding 100 m (330 ft).

BOTANY IN ACTION

Cycads in the garden

A small number of cycads are common in gardens, with the Japanese sago palm (*Cycas revoluta*) being the most common. Other genera include *Zamia* and *Macrozamia*. When buying, ensure that cycads come from a cultivated source (rather than from the wild), since all species are protected by the Convention on International Trade in Endangered Species (CITES).

Cycas revoluta, Japanese sago palm

Angiosperms: flowering plants

The flowering plants (angiosperms) are the largest and most diverse group of land plants. Like the gymnosperms, they are spermatophytes – seed-producing plants – but they form a distinct group by virtue of one main difference: they produce flowers.

Botanically, however, there are a number of other differences, and these include:

- Seeds are enclosed in a carpel, a unit of the ovary.

- The ovary develops considerably after fertilisation so that mature seeds are enclosed within a fruit. True fruits are a unique feature of flowering plants.

- The seeds contain a nutrient-rich material, called the endosperm, which provides food for the nascent plant.

In angiosperms, the gametophyte generation is so reduced that it comprises just a few cells within each flower, and this is shown in more detail in chapter 4 (p. 88). There are further details on the anatomy of angiosperm flowers, seeds and fruit in chapter 2.

Within the plant kingdom, flowering plants have undergone the greatest degree of evolutionary specialisation. The characteristics that differentiate them from other plants provide a number of ecological advantages that have ensured their success, enabling them to cover most of the land on the planet and survive where other plants cannot.

A fossil of an early flowering plant, *Dillhoffia cachensis*, now extinct, dated at 49.5 million years old.

Angiosperm ancestors

The ancestors of flowering plants developed from gymnosperms, and from the fossil record this is understood to have occurred between 245 and 202 million years ago. Gaps in the fossil record, however, make it hard for scientists to be sure of the precise details. It is probable that the earliest angiosperm ancestors were small trees or large shrubs adapted to growth in well-drained, hilly areas, not exploited by gymnosperms.

The first true angiosperms appear in the fossil record approximately 130 million years ago, with *Archaefructus liaoningensis* being the earliest-known angiosperm fossil. This species, like most of the other early angiosperms, are now extinct as they were rapidly replaced by more successful species, but ancient species do still exist in warm-temperate to tropical climates that have remained unchanged for millennia. The best example must be *Amborella trichopoda*, a rare shrub confined to New Caledonia in the Pacific.

Angiosperms began to take over habitats dominated by bryophytes and cycads from about

Magnolias are among the oldest flowering plants and their reproductive organs bear similarities to those of gymnosperms.

lilies), *Laurus* (laurel), *Drimys*, *Peperomia*, *Houttuynia* and *Asarum*. Unsurprisingly, if you look at the flower anatomy of some of these plants (a magnolia flower is a good example), there are clear comparisons with gymnosperms. For example, the stamens may be scale-like, resembling the male cone scales of a conifer, and the carpels are often found on a long flowering axis like a female gymnosperm cone.

The *Asteridae* are one of the most recent subclasses to have evolved and in it we see modifications of flower anatomy that maximise the efficiency of pollination and seed dispersal. For example, petals are commonly fused together and the flowers are often much reduced and grouped together (a sunflower is a good example of this, as it is basically an enormous flowerhead made up of hundreds of tiny flowers).

100 million years ago. By 60 million years ago they had largely replaced all gymnosperms as the dominant tree species. At this time, flowering plants were largely woody species, but the later emergence of herbaceous (non-woody) flowering plants led to another leap in angiosperm evolution. Herbaceous plants tend to have much shorter life spans than woody species, and so they are able to generate more variation in a smaller time frame, and thus evolve more quickly.

We also see during this period the emergence of a separate angiosperm lineage: the monocotyledons. The entire angiosperm class can be divided along these lines, with monocotyledonous plants making up about one-third of all flowering plants, and dicotyledonous plants making up the remaining two-thirds. (The differences are explained on p. 28.) It is estimated that the number of species of flowering plants currently in existence is in the region of 250,000–400,000.

Members of the subclass *Magnoliidae* are among the earliest flowering plants. This subclass contains a number of flowering plant genera that will be familiar to gardeners, such as *Magnolia*, *Nymphaea* (water

Tanacetum coccineum, pyrethrum, like most daisies, maximises pollination by grouping lots of small individual flowers (florets) together.

Flowering characteristics

Angiosperm flowers are highly varied. From catkins to umbels, grass flowers to orchids, to a simple buttercup, there are great differences and many specialisations, yet they all follow the same basic structure.

The perianth is a collective term used to describe external flower parts: the petals and sepals. Often the petals and the outer sepals are quite different, with the sepals being green and slightly leaf-like and the petals colourful and showy. In many instances, however, they are indistinguishable from each other (such as in tulips and daffodils, where they are sometimes called 'tepals'), and in other cases one or the other, or even both, may be entirely reduced or absent. In poppies (*Papaver*), the sepals encase the flower bud, but these rapidly drop off and will be absent from the open flower. In more advanced flower forms, parts of the perianth may be fused together. In strawberries (*Fragaria*), as with nearly all fruit-bearing plants, the petals fall as the fruit swells, leaving behind the leafy sepals that people often remove from the fruit before it is eaten.

The stamen is a term given to the male parts of a flower, made up of the anther and the filament. The filament supports the anther, and the anther bears the pollen grains. Some plants bear only male or female flowers (unisexual), as seen in the genus *Ilex* (holly); on a female plant these male parts are absent.

The pistil is the name for the female reproductive part of the flower. It is generally at the centre of the flower, surrounded by the stamens and then the perianth. It is made up of the stigma, the style and the carpel, or carpels. Within the carpel are the ovules. The stigma often has a sticky surface, as this is the landing point for grains of pollen, and it is often extended outwards by a long style so that it is in the best position to receive pollen. From the stigma, a successful pollen grain will grow a tube down through the style to reach an ovule, where fertilisation will take place.

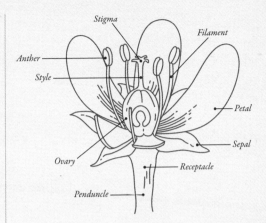

All flowering plants have evolved alongside the animals present in their environment, and the form that their flowers take is often a reflection of this. As a result, some very specific, and strange, pollination mechanisms, such as between bees and the bee orchid (*Ophrys apifera*) or between parasitic wasps and species of fig (*Ficus*). In such cases, the plants either deceive or offer some kind of benefit to the animal.

BOTANY IN ACTION

Flowering plants in the garden... and in agriculture

Angiosperms make up the majority of all ornamental plants, and along with the huge number of cultivars, they form a vast and extensive palette from which gardeners can work their art.

However, it must not be forgotten that agriculture is almost entirely dependent upon angiosperms, which provide virtually all plant-based foods as well as most feed for livestock. Of all the flowering plant families, grasses are by far the most economically important, providing many of the world's staples: barley, maize, oats, rice and wheat.

Monocotyledons versus Dicotyledons

About one-third of all flowering plants are monocotyledons. They are so called because their seeds possess only one cotyledon, or seed leaf. Dicotyledons possess two cotyledons, and when a seed germinates this difference can be readily seen. These names are often shortened to *monocot* and *dicot*.

Another important difference is the arrangement of their vascular bundles (the water-and-food transporting cells) within their main stems or trunks. In dicotyledons these are arranged cylindrically around the outer portion of the stem, which is why when you ring-bark a tree you will kill it, as you will be removing all these vital tissues. In monocotyledons, the vascular bundles are arranged randomly, so ring-barking is not possible.

In dicotyledons we also see primary tap roots, which do not occur in monocotyledons as the primary root soon dies off to be replaced by adventitious roots. We also see clear differences in leaf form: monocot leaves almost always have parallel veins, unlike the more complex vein networks seen on dicot leaves.

The flower parts of monocots are trimerous, meaning that they are arranged in multiples of three. Monocots also usually have well-developed underground structures, which are used as storage organs,

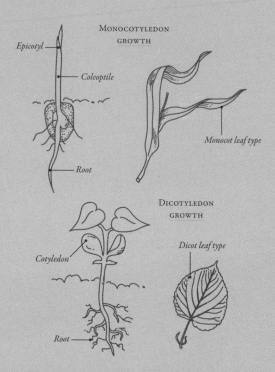

which the plants can draw upon during dormancy. The vast majority of monocots are herbaceous, although a few, like the palms, bamboos and yuccas, do put on woody growth. The arrangement of the vascular bundles means that the physical composition of their trunks or stems is quite different to that of dicotyledonous trees and shrubs.

Very few monocots are dominant in their habitats. The main exception here must be the grasses, which comprise one of the most successful plant groups ever, with over 10,000 species distributed over vast areas of the planet. One reason for their success must be due to their ability to withstand heavy grazing.

Crocuses, members of the *Iridaceae* family. Common garden bulbs, such as alliums, crocuses, daffodils and snowdrops, are all monocotyledons.

Plant naming and common names

Plant naming and the use of botanical Latin can be daunting to beginner gardeners. Nonetheless, the science of classifying living things – taxonomy – is essential to our understanding of the natural world. How else can you be entirely sure of which plant you are referring to and which plants you are growing and buying?

Taxonomy is the encyclopedia for biology, but living things constantly defy the artificial rules that scientists try to impose on them, and so any system of naming must allow for constant revision. Gardeners may be frustrated when the names of plants change, but as more is learnt and more species are discovered, the system must adapt.

Common names

The use of common or vernacular names seems an attractive proposition, as they are often easier to remember and pronounce. However, common names are often misapplied, misunderstood or get lost in translation from one language to another, which leads to great confusion and duplication. One problem is that common names differ from country to country, and even in different locations within the same country. Further complication arises when names are transliterated from non-Roman scripts, such as Japanese and Hebrew. The common name 'bluebell', for example, might refer to *Hyacinthoides non-scripta* in England, *Campanula rotundifolia* in Scotland, *Sollya heterophylla* in Australia and species of *Mertensia* in North America.

Most people do not think twice about using the names clematis, fuchsia, hosta, hydrangea and rhododendron as common names, yet these are also

Aquilegia vulgaris, columbine, granny's bonnet, American bluebell, granny's nightcap

the botanical names of these plants – they are far better known by their botanical name. When was the last time you heard someone referring to 'plantain lilies' – the common name of *Hosta*?

Common names can be misleading in other ways. For example, creeping zinnia is not a *Zinnia* (it is *Sanvitalia procumbens*), flowering maple (*Abutilon*) is not a maple (*Acer*), and evening primrose (*Oenothera*) is not a primrose (*Primula*). The scope for confusion is enormous, which is why we have botanical Latin.

Botanical names

The application of scientific names to plants is governed by a single set of rules, accepted and followed throughout the world: the *International Code of Nomenclature for Algae, Fungi and Plants* (ICN). Rules were also formulated for cultivated plants, resulting in the *International Code of Nomenclature for Cultivated Plants* (ICNCP), which governs the additional names sometimes given to plants in cultivation. All plant names must ideally conform to these two codes.

The origins of modern plant names

While the science of plant naming can be traced back to the Greek philosopher Theophrastus (370–287 BC), who wrote the very first manual of plant classification, it was not until the Renaissance that the taxonomy began its modern journey. Voyages of exploration at this time brought renewed interest in botany as rich new floras were discovered in faraway lands, such as tropical America.

Within the space of one hundred years, more than twenty times as many plants were introduced into Europe than in the preceding 2,000 years. Unable to rely on the works of Theophrastus, scientists were faced with the task of describing these species for themselves, for no previous records existed.

> 'WITHOUT NAMES, THERE CAN BE NO PERMANENT KNOWLEDGE'
> LINNAEUS

The first modern herbals were created by Otto Brunsfels and Leonard Fuchs in the 15th and 16th centuries, and in 1583 Andrea Caesalpino published *De Plantis libri XVI* (*16 Books on Plants*), one of the most significant works in the history of botany, noted for its careful observations and long descriptions in Latin. This was the first scientific treatment of flowering plants.

In 1596, Jean and Gaspard Bauhin published *Pinax Theatri Botanici*. In it, they named thousands of plant species and crucially shortened the Latin descriptions to just two words, thereby inventing binomial nomenclature: the two-word system of naming that botanists use today.

Species Plantarum (*The Species of Plants*), however, published in 1753, is generally regarded as the starting point for modern plant naming. It was written by the Swedish scientist Carolus Linnaeus, and in this and his later works he expanded upon the binomial system and created the academic discipline of defining groups of biological organisms on the basis of shared characteristics. Travellers, such as Joseph Banks on board HMS *Endeavour* with Captain Cook, sent him specimens from all over the world for classification.

Linnaeus also raised awareness of sexual reproduction in plants, since it was his observations of flower structure that formed the basis of much of his classification work. His *Systema Naturae* (*The System of Nature*) introduced the principles of Linnaean taxonomy and the consistent naming of organisms.

You may see author abbreviations used at the end of a plant's botanical name. This allows the name's authority to be traced back to its orginator. Linnaeus' abbreviation is 'L'.

BOTANY IN ACTION

Family, genus and species

For plants to be classified, and so make naming and reference easier, they are assigned to various taxonomic ranks. For gardeners, family, genus and species are the most important; any organism derives its name from the genus then species, in that order. The genus name should have an initial capital letter and the species epithet should be all in lower case; they should be styled in italics. Thus we have *Ribes nigrum* for blackcurrant.

Ribes nigrum, blackcurrant

Plant families

As with all levels of taxonomic classification, the aim of grouping plants into families is to make their study easier. Although families may appear at first to be of only academic interest for the gardener, knowledge of the family to which a plant belongs gives some indication of how it may look and perform in the garden. The names of plant families are written with an initial capital letter, and the ICN recommends that they are written in italics.

Cydonia oblonga, quince

Rosa chinensis 'Semperflorens', old crimson China rose

Family names

The names of the majority of families have traditionally been based on a genus within the family and have names ending in –*aceae*, such as *Rosaceae*, based on *Rosa*. However, there are a number of families that have not traditionally conformed to this pattern. While it is perfectly acceptable to continue to use these non-conforming names, the modern trend is to use names with -*aceae* endings. These families are as follows, with the more modern name in parentheses: *Compositae* (*Asteraceae*), *Cruciferae* (*Brassicaceae*), *Gramineae* (*Poaceae*), *Guttiferae* (*Clusiaceae*), *Labiatae* (*Lamiaceae*), *Leguminosae* (either *Fabaceae* or split into three families based on previous subfamilies: *Caesalpiniaceae*, *Mimosaceae* and *Papilionaceae*), *Palmae* (*Arecaceae*) and *Umbelliferae* (*Apiaceae*).

Barbara McClintock
1902–1992

Barbara McClintock, born Eleanor McClintock in Hartford, Connecticut, was an American scientist who became one of the world's most distinguished cytogeneticists, studying the genetics of maize.

She developed her passion for information and science during high school and went on to study botany at Cornell University's College of Agriculture. An interest in genetics and the new field of cytology – the study of cell structure, function and chemistry – led her to take the graduate genetics course at Cornell. And so began her life-long career in the development of maize cytogenetics and the study of the structure and function of the cell, especially the chromosomes.

She soon became recognised for her aptitude for, and thorough approach to, the subject. In her second year of graduate work, she improved on a method that her supervisor was using and she was able to identify maize chromosomes. It was a problem he had been working on for years!

In her groundbreaking cytogenetic work, Barbara studied maize chromosomes and how they change during the reproduction process. She developed the technique for visualising the maize chromosomes and, using microscopic analysis, demonstrated many fundamental genetic processes during reproduction, including genetic recombination and how chromosomes exchange genetic information.

She produced the first genetic map for maize, demonstrating that particular chromosome regions were responsible for creating a particular physical characteristic, and she demonstrated how the recombination of chromosomes correlated with new characteristics. Until this time, the theory that genetic recombination could happen during meiosis (see p. 88) was just that – a theory. She also showed how genes are responsible for turning physical characteristics on or off, and developed theories to explain the repression or expression of genetic information from one generation of maize to the next.

Unfortunately, Barbara was often thought of as being too independent and a bit of a 'maverick', not in keeping with most scientific institutes' ideas of a 'lady scientist'. As a result, she spent many years moving from institute to institute, especially between Cornell and the University of Missouri. She even spent some time working in Germany. Because her work on gene regulation was conceptually difficult to understand, it was not always accepted by her contemporaries. She often described the reception of her research as 'puzzlement, even hostility'. However, she was never deterred from continuing.

In 1936, she was finally offered a faculty position at the University of Missouri and was Assistant Professor for five years, until she realised that she would never be promoted. She left and worked for a summer at the Cold Spring Harbor Laboratory, finally accepting a

Barbara McClintock was one of the world's most distinguished cytogeneticists, renowned for her thorough approach and research.

Zea mays,
corn, maize

Dr Barbara McClintock determined that the variations in corn colour were due to specific genetic elements.

Because of her work with maize genetics, in 1957 she started research on the indigenous strains of maize found in Central and South America. She studied the evolution of maize races and how chromosomal changes had affected morphological and evolutionary characteristics in the plants. As a result of this study, Barbara and her colleagues published *The Chromosomal Constitution of Races of Maize*, which played a large part in the understanding of the ethnobotany, palaeobotany and evolutionary biology of maize.

Apart from the Nobel Prize, Barbara received numerous honours and recognition for her groundbreaking work, including being elected as a member of the National Academy of Sciences – only the third woman to be elected. She received the Kimber Genetics Award, the National Medal of Science, the Benjamin Franklin Medal for Distinguished Achievement in the Sciences, and was elected a Foreign Member of the Royal Society. She was also the first female President of the Genetics Society of America.

A small selection of some of the various corn mosaic colour variations that Barbara McClintock discovered were the result of chemicals inhibiting the synthesis of colour pigments.

full-time position the following year. It was at Cold Spring Harbor that McClintock figured out the process of gene expression in maize chromosomes.

For this and her other work, Barbara was awarded the Nobel Prize for Physiology or Medicine, the first woman to win the prize unshared. The Nobel Foundation credited her with discovering mobile genetic elements and she was compared to Gregor Mendel by the Swedish Academy of Sciences.

In 1944 she undertook the cytogenetic analysis of *Neurospora crassa*, a type of bread mould, at Stanford University. She successfully described this species' number of chromosomes and its entire lifecycle. *N. crassa* has since become a model species for classical genetic analysis.

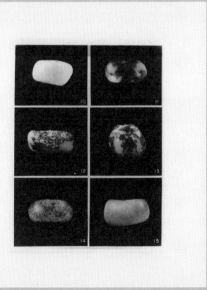

Genus

A genus (plural: genera) is a taxonomic group containing one or more species. Gardeners often use the word 'family' when 'genus' would be more appropriate. For example, it wouldn't be a surprise to hear a person claim that all apples belong to the same family. While this isn't technically wrong, the family actually encompasses the entire *Roseaceae*: pears, roses, geums, hawthorns, etc. What is probably meant is that they all belong to the same genus (*Malus*).

Species in the same genus share a number of significant physical attributes, making this possibly the most useful level at which plants can be identified for gardening, horticultural and practical purposes. In most genera the alliances are quite clear, as with *Geranium* for example, but with others the range of plants encompassed within the genus are so diverse that it can be difficult to imagine that one can be related to the other. Take for example the desert plant *Euphorbia virosa* and compare it with the popular garden *Euphorbia polychroma* – the differences are staggering, hardly surprising in a such a huge genus of over 2,000 species, yet their commonality lies in the flowers.

Ginkgo biloba,
maidenhair tree

Genera can vary in size from a single species to several thousand. For instance, *Ginkgo biloba* and the Venus flytrap, *Dionaea muscipula*, are the single representatives of their genus, whereas *Rhododendron* contains around 1,000 species.

Sometimes the genus name is abbreviated to a single letter followed by a dot. This is seen where the same genus of plant is mentioned repeatedly and there can be no confusion as to the meaning. Where the context is clear it is not necessary to spell out *Rhododendron* every time; it can be reduced to '*R*'.

Malus domestica,
'Crimson Beauty'

Malus is just one of the many genera in the very large plant family, *Rosaceae*, which includes *Crataegus*, *Rubus* and *Sorbus*.

Three Fascinating Genera

Viscum

The mistletoe genus contains between 70 and 100 species of woody, partially parasitic shrubs. They have a unique strategy of acquiring nutrients through a combination of their own photosynthetic activity and the absorption of materials from their host.

They are also known as 'obligate parasites' as they are unable to complete their life cycle without attachment to the host. These hosts are woody shrubs and trees, and different species of *Viscum* tend to parasitise particular host species, although most are adaptable to a number of different host species.

Passiflora

This climbing plant with its curious flowers is abundant in South America, and Spanish Christian missionaries took to using it to teach the story of Jesus, particularly the crucifixion, in a rather inventive way. The common name 'passionflower' reflects this, as it refers to the passion of Jesus:

- The climbing tendrils were used to describe the whips used in the flagellation.

- The ten petals and sepals were said to be ten of the twelve apostles (excluding Judas Iscariot, who betrayed Jesus, and St Peter, the denier).

- The ring of more than 100 filaments was said to represent the crown of thorns.

- There are three stigmas and five anthers: these were chosen to represent the three nails and the five wounds.

Passiflora caerulea, blue passionflower

Nepenthes

This genus of carnivorous pitcher plants shows how specialised plants can become in order to adapt to their environment. Their hugely modified leaves form jug-like traps to catch insects, and sometimes larger animals. These are digested within and used as a source of nutrients. Each leaf or 'trap' has three components: the lid, which may serve to prevent rainwater entering the pitcher and diluting the digestive juices; the colourful rim, which functions as a lure to insects; and the 'pitcher', which holds a liquid that attracts, drowns and finally digests its prey.

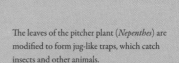

The leaves of the pitcher plant (*Nepenthes*) are modified to form jug-like traps, which catch insects and other animals.

Species

What are species? They are groups of interbreeding populations, which are reproductively isolated from other species.

The fundamental unit of plant classification is the species (not 'specie' – *species* is both singular and plural). The word is often abbreviated to 'sp.' singularly, or 'spp.' in the plural. Species can be defined as groups of individuals with many key characteristics in common, but distinct from other species within the same genus.

Plants within the same species form a set of interbreeding individuals that produce offspring with similar characteristics, and crucially they are almost always reproductively isolated from other species. Isolation is an important factor, as this allows for populations to evolve separate characteristics that are distinct from other similar populations.

The shifting of the continents over millions of years has caused much speciation, and thus we have plants such as *Platanus orientalis* in the eastern Mediterranean and *P. occidentalis* in eastern North America.

Platanus × hispanica,
London plane

This species is a hybrid (designated by the ×) between *Platanus orientalis* and *P. occidentalis*.

Platanus orientalis,
oriental plane

Both are types of plane tree. Ecological barriers also cause speciation; for example, *Silene vulgaris* is found inland and *S. uniflora* grows exclusively in a coastal environment.

Speciation, however, is not something that happens only over geological timescales. We see it happening even today, whenever isolation comes into play. An example of man-influenced ecological isolation is on land contaminated by copper waste in Wales – a legacy of the mining industry. On Parys Mountain in northeast Anglesey, copper pollution has destroyed much of the vegetation in affected areas; here scientists are beginning to observe speciation in copper-tolerant forms of the grass *Agrostis capillaris*.

When different species of the same genus that have been isolated for millennia are brought together we sometimes find that they can still breed together. In such cases we observe hybridisation, and the creation of a hybrid. Thus it is thought that the London plane tree (*P. × hispanica*) arose spontaneously when the two plane trees mentioned previously were grown together in Spain in the 17th century.

BOTANY IN ACTION

Species names

A species name can sometimes help gardeners learn more about the plant, as it may contain reference to how it grows, or where it naturally occurs. Some of the most commonly used species names include:

- *aurea, aureum, aureus*: golden-yellow
- *edulis*: edible
- *horizontalis*: growing horizontally or close to the ground
- *montana, montanum, montanus*: from mountains
- *multiflora, multiflorum, multiflorus*: many-flowered
- *occidentale, occidentalis*: relating to the west
- *orientale, orientalis*: relating to the east
- *perenne, perennis*: perennial
- *procumbens*: creeping or prostrate
- *rotundifolia, rotundifolium, rotundifolius*: round-leaved
- *sempervirens*: evergreen
- *sinense, sinensis*: from China
- *tenuis*: slender
- *vulgare, vulgaris*: common

Lifespan of species

Many plant species have a long lifespan, sometimes thousands of years, whereas others may only live for a matter of weeks. The oldest recorded plant is a Great Basin bristlecone pine, *Pinus longaeva*, in the White Mountains of California, USA, which was 5,063 years old in 2013. Several plants could contend for having the shortest lifespan, but the annual *Arabidopsis thaliana*, native to Europe, Asia and northwestern Africa, can complete its life cycle in six weeks.

Annuals complete their life cycle – germinate, grow, flower and set seed – all within one year or in one season, or in some plants much more quickly, as in the example above.

Biennials take two years to complete their life cycle. In the first year, they germinate and produce vegetative growth (leaves, stems, and roots), after which they become dormant, usually to survive a period of adverse weather or environmental conditions. The following year, usually after a period of further vegetative growth, they flower, set seed and then die.

Perennials live for more than two years. Like biennials, they may become dormant from year to year, sometimes dying back below ground, but they come back into growth during the favourable seasons and usually flower and fruit each year. To a gardener, the term 'perennial' usually means a herbaceous perennial; any non-woody plant that a botanist would call a herb (see chapter 2, p. 47). To a gardener, a herb is a totally different thing altogether – subtle differences exist between the jargon of gardeners and botanists.

Tulipa montana (mountain tulip); the name refers to its natural habitat in the stony hills and mountains of northwest Iran, not the actual state of Montana in USA.

Subspecies, varieties and forms

There can often be huge variation within a species. Botanists deal with these anomalies by arranging some species into subspecies, varieties and forms. As with all levels of taxonomic classifications, rules exist for their naming.

With some species, such as the monkey puzzle tree (*Araucaria araucana*), which has a very limited natural range, little variation is seen among individuals. It is only ever known by its species name, therefore, as no variant exists, plant breeders have not created any cultivated forms.

Some other species, however, are very complicated. *Cyclamen hederifolium* (autumn cyclamen) is a great example. There are at least two varieties (*confusum* and *hederifolium*), and one of these varieties is split into two forms (*albiflorum* and *hederifolium*).

Subspecies

When wild plants have a wide distribution, populations may acquire slightly different characters in different geographical areas, especially if isolated. Such populations may be distinguished as subspecies (abbreviated to 'subsp.' or 'ssp.') within the one species.

Thus we see two different subspecies of *Euphorbia characias* within its natural range in southern Europe: subsp. *characias* inhabits the western Mediterranean and subsp. *wulfenii* the eastern Mediterranean, and they differ in plant height and colour of the nectaries.

Araucaria araucana,
monkey puzzle tree

The monkey puzzle tree is only known as the species and does not have any forms or varieties.

Variety

Distinct populations and individuals occurring sporadically within the geographical range of a species or subspecies may be recognised as varieties (technically *varietas*, abbreviated to 'var.') They may occur throughout the range of a species or subspecies, but are often not correlated with geographical distribution. An example is *Pieris formosa* var. *forrestii*.

Varieties are naturally occurring and they will usually hybridise freely with other varieties of the same species. The word 'variety' is often wrongly used to describe a cultivar (see opposite page). Unlike varieties, which are usually naturally occurring, cultivar names are only applied to varieties bred in cultivation or of cultivated origin.

Form

A form (technically *forma*, abbreviated to 'f.') is the lowest botanical subdivision in common usage. It is used to describe minor but noticeable variations, such as in flower colour, and as with variety it is not usually correlated with geographical distribution.

Two examples are *Geranium maculatum* f. *albiflorum*, a form of *G. maculatum* with white flowers, and *Paeonia delavayi* var. *delavayi* f. *lutea*, a form of tree peony with yellow flowers.

Hybrids and cultivars

Gardeners will need to become familiar with the terms 'hybrid' and 'cultivar' as they are commonly used in the names of ornamental plants – improved forms that result from the activities of plant breeders.

Hybrids

Some plant species, when growing together in the wild or in a garden, will interbreed either naturally or by means of human intervention. The resulting offspring are known as hybrids. A multiplication sign in their name indicates hybrid status, for example *Geranium* × *oxonianum*, a hybrid between *G. endressii* and *G. versicolor*. Hybrids between different species within the same genus are known as interspecific hybrids and those between different genera are known as intergeneric hybrids.

The majority of hybrids occur between species within the same genus. For example, hybrids between the heathers *Erica carnea* and *E. erigena* are given the hybrid name *Erica* × *darleyensis*.

Hybrids between different genera are given a new hybrid genus name, and the different combinations of species are treated as species in their own right. Here the multiplication sign is placed before the genus name. The hybrid between *Mahonia aquifolium* and *Berberis sargentiana* is called × *Mahoberberis aquisargentii*, and that between *Rhodohypoxis baurii* and *Hypoxis parvula* is called × *Rhodoxis hybrida*.

There are also a few special hybrids, called graft hybrids or graft chimeras, where the tissues of two plants are physically rather than genetically mixed. These are indicated by a + sign. So *Cytisus purpureus* and *Laburnum anagyroides* produce + *Laburnocytisus* 'Adamii', and graft chimeras between *Crataegus* and *Mespilus* produce + *Crataegomespilus*.

Cultivars

In cultivation, new plants with variations created or encouraged by plant breeders are called cultivars (a CULTIvated VARiety, sometimes abbreviated as 'cv'). They are given a cultivar name. To make it stand out from the purely botanical part of a name, the cultivar name is enclosed in single quotes and is not written in italics, as in *Erica carnea* 'Ann Sparkes'.

New cultivar names produced since 1959 now have to follow international rules and must, in part at least, be in a modern language to ensure they are distinct from the botanical part of the name. Previously, many cultivars were given names styled in Latin, such as *Taxus baccata* 'Elegantissima'.

Taxus baccata, English yew

Groups, grex and series

Where a lot of similar cultivars exist of the same species, or where a recognised cultivar has become variable (usually through poor selection of propagation material), they are sometimes given a more collective name, the Group name. These names always include the word Group and, where used along with a cultivar name, are enclosed in round brackets. For example, *Actaea simplex* (Atropurpurea Group) 'Brunette' and *Brachyglottis* (Dunedin Group) 'Sunshine'. The botanical ranks species, subspecies, variety and form may also be treated as Groups. This is useful where the rank is no longer recognised botanically, though still worth distinguishing in gardens, such as *Rhododendron campylogynum* Myrtilloides Group, where var. *myrtilloides* is no longer recognised botanically.

Brachyglottis, tree daisy

Actaea simplex, baneberry, bugbane

In orchids, where complex hybrid parentages are carefully recorded, the Group system is further refined. Each hybrid is given a grex name (Latin for 'flock'), which covers all offspring from that particular cross, however different they may be from one another. Grex is abbreviated to 'gx', so for example we have *Pleione* Shantung gx, and *Pleione* Shantung gx 'Muriel Harberd', which is a named cultivar selected from that grex. The grex name is written in Roman type (non-italic) without inverted commas, and gx follows rather than precedes the name.

Series names are often used with seed-raised plants, particularly F1 hybrids. They are similar to Groups in that they may contain a number of similar cultivars, but they are not governed by any nomenclatural code and are mainly used as marketing devices. The identities of individual cultivars are often not disclosed by the plant raisers, and the same colours may reappear over the years under a variety of names, to suit different markets.

Trade designations

In addition to a cultivar name, many cultivated plants are given 'selling names' or 'marketing names', which are referred to as trade designations. These are not regulated, are often different in different countries, and are sometimes replaced after a time. Some are also registered trademarks, which debars them from being treated as cultivar names.

They are often used where the cultivar name is a code or nonsense name, used only to register a plant for Plant Breeders' Rights (see below), that is impractical for commercial use. This is common practice with rose names. The cultivar name should be clear on the plant label but often takes second place to the trade designation. The rose with the trade designation Brother Cadfael has the cultivar name 'Ausglobe' and *Rosa* Golden Wedding is 'Arokris'. While the cultivar name remains constant (which is of use to gardeners, so they can be sure which plant they are buying), the trade designation can be varied to suit marketing needs.

Trade designations may resemble cultivar epithets and are often erroneously presented as such. They should not be enclosed in single quotation marks and should be printed in a typeface that contrasts with the cultivar epithet. For instance, *Choisya ternata* 'Lich' is marketed under the trade designation Sundance and should be written as *Choisya ternata* 'Lich' Sundance or *Choisya ternata* Sundance ('Lich'). It should not be written as *Choisya ternata* 'Sundance'.

Another type of trade designation derives from cultivar names originating in foreign languages. These trade designations may arise if the cultivar name given in a foreign language is found to be difficult to pronounce or read. An example is *Hamamelis* × *intermedia* Magic Fire ('Feuerzauber').

Plant Breeders' Rights (PBR)

These are a form of intellectual property legislation designed specifically to protect new cultivars of plants. They allow the breeder to register a plant as their own property if it meets certain internationally agreed criteria. It is designed to help plant breeders benefit commercially from their work.

If a PBR is granted, it covers specific territories for defined periods of time. The owner may then license companies to grow their cultivar and collect a royalty on each plant grown. PBR offices exist all over the world, and plants with PBR cannot be propagated for commercial purposes by any method – including seeds, cuttings or micropropagation – without permission from the owner.

Hamamelis × *intermedia* Magic Fire *('Feuerzauber'),* witch hazel

The trade designation Magic Fire arises from a loose translation of the German name, Feuerzauber.

Prunus domestica, plum

Chapter 2

Growth, Form and Function

Above a certain evolutionary level, all plants are made up of a complex system of specialised organs which carry out the different functions needed for the plant to live, grow and reproduce. The form these organs take is influenced by the habitat in which the plant evolved. Form, habitat and function, therefore, are all interconnected.

In growth, we would expect every part of a plant that performs some kind of function – such as succulent stems or hairy leaves – to make the plant more efficient in its native habitat.

The function of some characteristics is still not properly understood. Some parts of a plant's anatomy may even be redundant, having once had a function in a plant's evolutionary history, but no longer required in modern times.

Under cultivation, a plant's form and function probably have little to do with its habitat. Since the dawn of agriculture, humans have separated plants from the environmental pressures of their natural habitats and introduced pressures of their own, in order to cultivate and improve them. The end result is plants with forms and functions that suit the needs of man – either ornamental or productive.

Plant growth and development

Plants will grow when certain requirements are met: light, temperature and moisture. Different plants have different needs, and any or all three of these can trigger growth in particular plants. There are those that respond to day length (clover, for example, responds to increasingly longer days in spring), those that will only grow after rain (such as the resurrection plant from the Chihuahua desert, North America, *Selaginella lepidophylla*), or trees that will only come into leaf once the average air temperature rises above a certain level (such as the English oak, *Quercus robur*). Conversely, plants may stop growing when these conditions are not met, and in such instances dormancy (or perhaps death) may ensue.

Anastatica hierochuntica, the rose of Jericho

Meristematic tissue

All growth occurs at a cellular level, and in plants this is restricted to areas of meristematic tissue, where there are rapidly dividing cells. Growth here brings about changes in the overall form and structure of the plant, and ultimately an increase in the level of complexity of the organism.

These meristematic regions are 'initial cells' in that they are undifferentiated into any specialised form. They are only found in certain areas of the plant: such as at the root tips and in buds, the root and shoot tips (apical meristems) and the cambium layer immediately below the bark of a woody plant (lateral meristems), which allows stems or roots to increase in girth. The growth of the meristem is potentially indeterminate as it can continue for as long as the plant is alive.

Although an amazing variety of forms is produced by growth and development, they are all accounted for by just three events at the cellular level, and they all arise from the meristem:

1. Cell division (mitosis)

2. Cell enlargement

3. Cell differentiation (in which a cell, once it has reached its final size, becomes specialised and ceases to be meristematic)

The different ways in which these three steps can occur account for the variety of tissues and organs in individual plants, as well as for the different kinds of plants.

Newly formed meristems often enlarge in three dimensions, but in elongated plant organs, such as stems and roots, the enlargement occurs mostly in one dimension, with enlargement soon becoming elongation, owing to physical pressures on the enlarging cells. Such enlarged cells are laid down behind or beyond the meristem and extension growth takes place.

Intercalary meristems are a feature found exclusively in monocot plants (see chapter 1, p. 28), such as grasses (including bamboos). These are areas of meristematic tissue that lie at the base of nodes (in bamboo) or at the base of leaf blades (as in herbaceous grasses). This is most likely to be an adaptation in response to grazing by herbivores: think of pandas feasting on bamboo leaves, or cows chewing grass. It allows these plants to grow rapidly after grazing. To gardeners, this is of special relevance, every time they mow the lawn.

Apical meristems, or growing tips, are found in the buds shoot tips and at the ends of roots. They start the growth of new cells and so produce the developing root or shoot behind them.

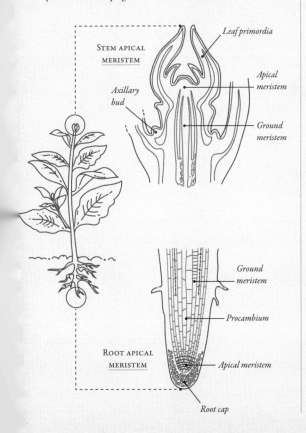

Cell differentiation

Once the meristematic cells have elongated, they begin to differentiate and mature into specialised cells.

Parenchyma tissue

Parenchyma tissue is perhaps the most prolific, made up of thin-walled 'general-purpose' plant cells with many varied functions. Usefully, they are potentially meristematic, meaning they can revert to meristematic form should the need arise.

The main, general cells are called parenchyma.

Cross-section

Longitudinal section

BOTANY IN ACTION

Functions of parenchyma tissue

Parenchyma tissue is plentiful, bulking out the volume of a plant, but besides this use as a 'packing material' it serves a number of useful functions:

- **Water storage:** In plants that exist in dry areas, water reserves can be held in the parenchyma tissue, which also serves to stiffen the plant.

- **Aeration:** The large spaces between the parenchyma cells can be used to facilitate gas exchange, and air spaces can be useful for some aquatic plants that require the extra buoyancy.

- **Food storage:** Parenchyma cells can be used to store food in solid form or in solution as sap.

- **Absorption:** Some parenchyma cells are modified to take up water from the surrounding environment for the plant's needs (e.g. in root-hair cells).

- **Protection:** Some parenchyma tissue helps to prevent damage to the plant.

Collenchyma tissue

Collenchyma tissue is similar to the parenchyma, but with its cell walls thickened by cellulose. It therefore functions as supportive tissue, especially in young stems and leaves, but it is flexible enough to accommodate extra growth. Collenchyma can revert back to meristematic tissue, to form lateral meristems.

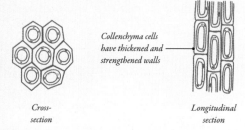

Collenchyma cells have thickened and strengthened walls

Cross-section

Longitudinal section

Sclerenchyma tissue

Sclerenchyma tissue has greatly thickened walls impregnated with lignin (woody material). This tissue only matures when growth is complete, as once the lignin is deposited it cannot accommodate further growth and at this point the cell dies, leaving just the woody walls. It is the sclerenchyma tissue that mechanically supports woody plants.

Two basic forms of sclerenchyma exist: fibres and sclereids. Fibres are long cells with tapered ends, and are often grouped into bundles. Sclereids are spherical cells, and they are frequently encountered in the flesh of succulent fruit. It is sclereids that give pears (*Pyrus*) their gritty texture.

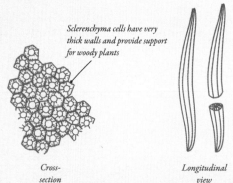

Sclerenchyma cells have very thick walls and provide support for woody plants

Cross-section

Longitudinal view

Plant types: botanists v. gardeners

Most gardeners probably divide plants up into ten or eleven broad categories depending on their ornamental use: trees; shrubs; climbers; herbaceous perennials; rock plants (alpines); annuals and biennials; bulbs, corms and tubers; cacti and succulents; ferns; herbs and aquatic plants. Other groups that might also be included are cycads and mosses.

Botanists look at the world differently. While they may use some or all of these terms, they often have strict definitions that may seem strange or counter-intuitive to the gardener. Some of the most commonly used terms are defined below.

Trees and shrubs

Botanists may simply refer to any tree or shrub as a woody plant, since there is no clear distinction between a tree and a shrub. However, they may also use the term 'phanerophyte', which applies to any plant that projects itself into the air on stems, with its resting buds borne more than 25 cm (10 in) above the ground.

Some woody plants, such as sweet chestnut (*Castanea sativa*), are clearly trees, but others blur the lines, such as lilac (*Syringa*) or holly (*Ilex*), which show both a shrubby and a tree-like habit depending on the species or cultivar, or the way it is grown. The term 'mallee' will be familiar to Australians as it is used to describe eucalyptus trees with a shrubby, multi-stemmed habit. Botanists may use the term 'chamaephyte' to describe subshrubs – those with resting buds no more than 25 cm above the ground.

Gardeners usually say that a shrub must have multiple trunks, whereas only a tree can have a single trunk. Various horticultural authors have tried to define a tree, and most say that it should be capable of exceeding a certain height on a single stem – some say 3 m (10 ft), others 6 m (20 ft) – sometimes adding

Syringa vulgaris,
lilac

that the stem should be able to exceed 20 cm (8 in) in diameter. As is to be expected, this is a grey area, and we have to live with the fact that some small trees may, in some people's eyes, simply be large shrubs.

Perennials, herbaceous perennials and herbs

Any plant that lives for more than two years is known as a perennial, which includes any tree, shrub, bulb, rhizome or alpine plant. Gardeners may baulk at this hijacking of a term they reserve exclusively for all the non-woody, mostly flowering border plants that die down in winter, but in this case you can't argue with the botanists: the dictionary defines the word 'perennial' as 'constantly recurring or lasting for an indefinite time'.

So for this reason, gardeners are better off sticking to the term 'herbaceous perennial' to define their border plants. Such a term excludes all woody plants (although the line is blurred when we come to some plants like salvias, thymes or perovskias, which put on a little woody growth and are sometimes regarded as subshrubs), and it also excludes bulbs, corms and tubers, alpine plants and cacti. The word 'herbaceous', however, presents further difficulty in that it botanically applies to plants whose top-growth dies off completely during the plant's dormant season (usually winter). The gardener will know that many herbaceous perennials, however, such as hellebores and some ferns (such as the hart's-tongue fern, *Asplenium scolopendrium*) do not die back – they just keep on going. In these cases, tired foliage can be snipped off to maintain a fresh display.

Botanists create further confusion when they talk of 'herbs'. We all know herbs as handy culinary plants that grow in the kitchen garden, such as rosemary and basil (*Ocimum basilicum*). They also include plants with herbal or medicinal uses, such as lavender and evening primrose (*Oenothera*). To a botanist, a herb is any non-woody plant – a term that can cover a whole multitude of plants. The term 'hemicryptophyte' may also be used, which describes a plant with its resting buds at or near the soil surface.

Asplenium scolopendrium,
hart's-tongue fern

> ## BOTANY IN ACTION
>
> One fascinating group of plants that gardeners are unlikely to have come across are the megaherbs. They are like giant herbaceous perennials and are confined to the subantarctic islands off the coast of New Zealand. Plants that evolve in isolation like this often show unique characteristics, and these are no exception: the Campbell Island daisy (*Pleurophyllum speciosum*), for example, forms enormous rosettes up to 1.2 m (4 ft) across. Giant herbs are found in isolated communities, such as the giant groundsels (*Dendrosenecio*) of Mount Kilimanjaro. Gardeners may fancy that some ornamental herbaceous perennials warrant 'megaherb' status, such as the globe artichoke (*Cynara cardunculus*) and the huge grass *Miscanthus* × *giganteus*.

Perennial plants that flower and fruit only once in their lifetime and then die are termed 'monocarpic'. Examples include some yuccas and bamboos, and some *Meconopsis* and *Echium*. Most perennials, however, flower over many years during their lifetime, usually every year, and they are known as 'polycarpic'.

Annuals and biennials

Plants that gardeners know as annuals are categorised by botanists as therophytes. They flower, set seed and then die all within the favourable growing season, enduring as seeds during their dormant phase. Hence the term 'annual', which is self-explanatory. Gardeners will be very familiar with annuals as they can bring a quick splash of colour to the garden, and commonly grown species include the poached-egg plant (*Limnanthes douglasii*), love-in-a-mist (*Nigella damascena*) and nasturtiums (*Tropaeolum majus*).

Confusion sets in when gardeners talk of bedding plants. Often these are described as annuals, because they are put in place for just one or two seasons and then discarded. Bedding plants are commonly perennials, such as begonias and busy Lizzies (*Impatiens*), but because they will not overwinter in frost-prone climates (or begin to grow tired and tatty after one season) they are grown as annuals.

Biennials are plants that live for just two years, and then flower, set seed and die – they are monocarpic. Foxgloves (*Digitalis purpurea*) and pride of Madeira (*Echium candicans*) are two well-known examples. Many vegetables are biennials. Some perennials with short lifespans, such as wallflowers (*Erysimum*), are sometimes classed as biennials, although once again the lines become blurred. The fact that some herbaceous perennials live longer than others can be a bugbear for gardeners, who may be stumped by the sudden death of one of their favourite plants. The fact is that some perennials simply have longer lives than others: echinaceas can be relatively short-lived; irises may simply go on and on.

Tropaeolum majus, nasturtium

Climbing plants

Clematis vitalba,
old man's beard, traveller's joy

Botanists may refer to climbing plants as vines. Very big tree climbers seen in the tropics, such as the tortoise's ladder (*Bauhinia guianensis*), are known as lianas, although large climbers in temperate forests, such as old man's beard (*Clematis vitalba*) could also be defined as lianas.

Gardeners may prefer to use the term 'climber' instead, reserving the word 'vine' for grapevines (*Vitis*). Climbers can be herbaceous, such as sweet peas (*Lathyrus odoratus*), or woody, such as wisterias and grapevines.

Various types of climbers are described by gardeners: scramblers, twiners and clingers are just three examples. Others may sprawl, arch or trail. These various terms all loosely describe the manner of growth that the climbing plant shows, yet they all have one thing in common, and that is that they are most likely to be too soft or too weak to fully support themselves. Their lifestyle means that they need to use nearby plants as a support so that they can reach the light and present their flowers and fruit to the world. The gardener may grow climbers through other plants (such as clematis through roses) or against a support such as trellis.

Clingers

These cling to whatever they are growing on by means of adventitious aerial roots or self-adhesive sucker pads produced at the end of tendrils. Examples of plants that produce aerial roots include ivy (*Hedera*) and some species of climbing hydrangea. They do not need any supplementary support structures in the garden to grow on – they will naturally climb whatever they are placed against.

Twiners

Twiners are those plants that climb by producing curling or twining leaf tendrils, leaf stalks or stems, which entwine the support. Examples include *Clematis*, *Lonicera* (honeysuckle) and *Wisteria*. A supplementary support system, such as trellis, wires or mesh, is needed in the garden.

Scramblers

Scramblers gain purchase and clamber up their supports by using hooked thorns (as with roses) or by rapid elongation of their shoots, such as the potato vine (*Solanum crispum*). In order to keep them secure, a strong support system is needed in the garden onto which the stems can be tied.

There are a number of shrubs that are grown against walls and fences in gardens, although they do not naturally climb. Examples include *Pyracantha* and *Chaenomeles*, and many fruit trees can also be grown in this manner. This allows gardeners to make the most of all the available space in their gardens, as the plants can be encouraged to grow in just two dimensions.

In evolutionary terms, therefore, trees and shrubs that host climbers are at a distinct disadvantage. Gardeners may notice that eucalyptus trees in their garden go through an annual process of shedding their bark. It has been suggested that this adaptation makes it harder for climbing plants to take hold.

Bulbs, corms and tubers

Collectively known as 'cryptophytes' by botanists, a catch-all term for any plant that grows from an underground structure, such as with tulips (*Tulipa*). It includes those that grow in dry soil (geophytes), those that grow in marshy soil (helophytes) and those that grow underwater (hydrophytes – such as waterlilies).

It is a shame that 'cryptophyte' is not a more widely used term, since it would be useful to most gardeners who find the subtle distinction between what is a bulb, corm, rhizome or tuber rather confusing and a little nonsensical. As it is, gardeners are stuck for the time being with these four terms.

Bulbs are an underground type of bud with a very short, thick stem and tightly packed, fleshy scales. An onion is a typical bulb. Corms are swollen underground stems without scales, as in *Crocus* and *Gladiolus*. Tubers are underground storage organs that can be separated from the main plant and regrown; in cold climates tubers are often lifted and overwintered in a frost-free place. Dahlias and potatoes are tuber-producing plants. Rhizomes are horizontally creeping stems lying just on the surface of the soil or partially underground. Bearded irises are typical rhizomes. Bulbs, corms, tubers and rhizomes are discussed in further detail on pp. 82–83.

Rock plants

This group of plants refers more specifically to the type of habitat that the plant comes from rather than the type of plant it is. Rock plants could be, for example, shrubs, herbaceous perennials or annuals. They are sometimes known as alpine plants. The one thing that they all have in common is that they come from cold habitats with a short growing season, low winter rainfall (with the moisture locked up as snow), freezing winter temperatures and often well-drained soils.

High mountain pastures or rocky screes are a typical habitat, but plants enjoying similar conditions such as coastal cliffs and shores might also be included. Alpine habitats are typically at the tops of mountains, above the tree line. The closer one gets to the poles, the closer to sea level these habitats get. In the tropics, alpine zones are only seen at the peaks of the highest mountains. Gardeners may build special alpine houses, sharply drained rock gardens or scree beds and dry stone walls to grow these plants and provide them with the exacting conditions that they need.

BOTANY IN ACTION

Epiphytes

This term is used to describe any plant that grows on another plant (usually a tree), or sometimes on man-made structures, without drawing any nutrients from its host. Its roots serve to hold the plant in position and will draw moisture from the air or from the surface of the host. Decaying matter that collects in cracks and crevices will supply nutrients, and the epiphytes themselves may have special structures to collect rainwater.

Epiphytes are not common in temperate gardens. In subtropical or tropical areas they are more evident, such as the staghorn ferns (*Platycerium*) and Spanish mosses (*Tillandsia*). A huge number of orchids are epiphytic, which accounts for their strange rooting requirements when grown as houseplants.

Tillandsia, air plant

Tillandsias are epiphytic plants belonging to the bromeliad family. They are also referred to as 'air plants'.

Buds

A bud is defined as a dormant projection on a stem from which growth may develop when the environmental conditions are favourable. This growth will either be vegetative or floral, depending on the type of bud, although some buds can produce root growth.

Buds are usually produced in the leaf axil or at the tip of the stem, but it is not unusual to find them on other parts of the plant. They can remain dormant for some length of time, only becoming active when needed for growth, or they may grow as soon as they are formed.

Bud morphology

Scaly buds

Scaly buds are found on many plants from cooler climates, and they are named for their protective covering of modified leaves, called scales. These tightly enclose the delicate parts of the bud below them. Bud scales may also be covered by a sticky gum, which provides further protection to the bud. With deciduous trees, it is often possible to identify a species of tree by the shape of its buds and the number of bud scales.

Plants produce numerous types of buds, which can be classified by where they occur on the plant, how they appear and their function.

By location	By status	By morphology	By function
Terminal	Accessory	Scaly	Vegetative
Axillary	Pseudoterminal	Covered	Reproductive
Adventitious	Dormant	Hairy / Naked	Mixed bud

Naked buds

Naked buds are those without scale protection, having instead a covering of tiny, undeveloped leaves, which are often very hairy. These hairs provide some protection, and are sometimes referred to as hairy buds. Some scaly buds also have this hairy protection, such as with pussy willow flower buds (some *Salix* species).

Many annual and herbaceous plants produce no obvious buds. In fact, in these plants the buds are very reduced, simply consisting of a mass of undifferentiated meristematic cells in the leaf axils. Conversely, the buds of some herbaceous plants are not thought of as buds, although that is what they are. For example, the edible parts of some vegetables in the cabbage family (*Brassicaceae*) are buds: a cabbage head or heart is simply an enormous terminal bud, Brussels sprouts are large lateral buds, and cauliflower and broccoli heads are formed of masses of flower buds.

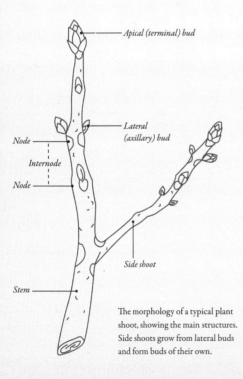

The morphology of a typical plant shoot, showing the main structures. Side shoots grow from lateral buds and form buds of their own.

The different types of buds

Within a single plant, there will be various types of buds, all with different functions.

Apical or terminal buds

Apical or terminal buds appear at the tip of a stem, and their growth exerts a level of control over the lateral buds lower down, owing to its production of growth-regulating hormones. This is called apical dominance, and is seen most strongly in some conifers. The Nordmann fir (*Abies nordmanniana*), often grown as a Christmas tree, is a good example with its strongly pyramidal shape. If an apical bud is lost, through grazing or other damage, dominance is lost and buds lower down will start to grow more strongly, in an effort to replace the apical bud. Gardeners may take advantage of this response to encourage bushy growth.

Lateral or axillary buds

Lateral or axillary buds are usually formed in the leaf axils – the point at which the leaf joins the stem. They will usually develop into leaves or side shoots.

Adventitious buds

Adventitious buds occur in other places on the plant, such as on the trunk, leaves or roots.

Some adventitious buds form on roots, producing new growth that gardeners call 'suckers'. These are seen on many species of tree, such as the stag's-horn sumach (*Rhus typhina*), and can become a nuisance, although they are sometimes used as a method of propagation.

Adventitious buds that form on the leaves of some plants may also be used for propagation, which is useful to both the plant and the gardener. In the garden, the thousand mothers plant (*Tolmiea menziesii*) makes a rapidly spreading groundcover plant precisely because of this ability.

GROWTH, FORM AND FUNCTION

Eucalyptus obliqua,
messmate, Australian oak

Epicormic buds

Epicormic buds are a type of adventitious bud. They lie in a dormant state under the bark of some woody plants. Their growth can be triggered in response to physical damage, or when there are no other buds left. Gardeners usually call upon epicormic buds when they perform hard pruning, but it must be noted that not all plants can be relied upon to produce shoots from epicormic buds. Many conifers, for example, as well as lavender and rosemary, can be killed by hard pruning.

Eucalyptus trees in Australia rely heavily on epicormic buds, which is an evolutionary adaptation to bush fires. In these trees the buds are set very deep in order to resist extreme temperatures. Growth of these dormant buds is triggered by fire, after which regenerative growth begins.

Some species are able to grow roots from adventitious buds, and gardeners exploit this ability when they take cuttings. Willows and poplars are able to take root easily from bare stems cut from the plant in winter (hardwood cuttings), and roses can also be propagated this way. In some trees, several epicormic buds break at the same point, producing a profusion of thin stems called 'water shoots'.

Vegetative and flower buds

Vegetative buds, or leaf buds, are usually small and thin and go on to produce leaves. Reproductive or flower buds, sometimes referred to as fruit buds in fruiting plants, are fatter and contain the embryonic flowers. Mixed buds contain both embryonic leaves and flowers.

The production of flower buds in fruit trees is a complex process and is determined by cultivar, rootstock, light levels, nutrients and water availability. It is, however, possible to influence this to some extent and promote flower bud production over the formation of vegetative buds – useful to the fruit grower trying to increase yields.

This is achieved in various ways, through careful pruning and by meeting the plant's nutritional requirements. The traditional forms of fruit training (fans, cordons and espaliers) also play a role as they serve to restrict the flow of sap, and so nutrients and hormones, by reducing vertical growth. These techniques are found to encourage fruit bud formation and reduce vegetative growth.

Rosa pendulina,
alpine rose

Many rose species only flower once, whereas modern cultivars produce flower buds throughout the summer.

Robert Fortune
1812–1880

Our gardens would be much the poorer if it weren't for the brave exploits of the enigmatic and notoriously surly botanist and plant hunter Robert Fortune. During several visits – mainly to China, but also Indonesia, Japan, and the Philippines – Fortune brought back more than 200 ornamental plants. These were mainly trees and shrubs, but they also included climbers and herbaceous perennials.

He was born in Kelloe, in what is now County Durham in the northeast of England, and was first employed in the Royal Botanic Garden, Edinburgh. Later he was appointed as Deputy Superintendent of the Hothouse Department in the garden of the Horticultural Society of London (later to be renamed the Royal Horticultural Society) in Chiswick. A few months later, Fortune was granted the position of the Society's collector in China.

He was sent on his first journey in 1843 with little pay and a request to 'collect seeds and plants of ornamental or useful kind, not already cultivated in Britain', as well as to obtain information on Chinese gardening and agriculture. He was especially tasked to find any blue-flowered peonies and to investigate the peaches growing in the Emperor's private garden, among other things.

Each trip enriched Britain's gardens and greenhouses with plants covering nearly the whole A–Z of genera from *Abelia chinensis* to *Wisteria sinensis*, including *Camellia reticulata*, chrysanthemums, *Cryptomeria japonica*, various *Daphne* species, *Deutzia scabra*, *Jasminum officinale*, *Primula japonica* and various *Rhododendron* species.

Although his travels resulted in the introduction to Europe of many new and exotic plants, probably his most famous accomplishment was the successful transporting of tea plants from China to the Darjeeling region of India in 1848 on behalf of the British East India Company. Fortune used one of the latest inventions to transport the plants – Nathaniel Bagshaw Ward's Wardian case. Unfortunately, most of the 20,000 tea plants and seedlings

A great explorer of the Far East in the mid-1800s, Robert Fortune returned to Britain with over 200 species of ornamental plant.

'The art of dwarfing trees, as commonly practised both in China and Japan, is in reality very simple... It is based upon one of the commonest principles of vegetable physiology. Anything which has a tendency to check or retard the flow of the sap in trees, also prevents, to a certain extent, the formation of wood and leaves.'
Robert Fortune in *Three Years' Wanderings in the Northern Provinces of China*

Camellia sinensis,
tea tree

Robert Fortune was instrumental in introducing tea to India from China, establishing the Indian tea industry that we know today.

Rhododendron fortunei,
rhododendron

Fortune found this plant growing at 900 m (3,000 ft) in the mountains of eastern China. It was the first Chinese rhododendron introduced to Britain.

perished, but the group of trained Chinese tea workers who came back with him, and their technology and knowledge, were probably instrumental in the setting up and the success of the Indian tea industry.

Fortune was generally well received on his travels, but did experience hostility and was once threatened at knifepoint by an angry mob. He also survived killer storms in the Yellow Sea and pirate attacks on the Yangtse River.

He became proficient enough at speaking Mandarin that he was able to adopt the local dress and move among the Chinese people largely unnoticed. This enabled him to visit parts of the country that were off limits to foreigners. By shaving part of his head and growing a ponytail, he was able to effectively blend in.

The incidents of his travels were related in a succession of books, which include *Three Years' Wanderings in the Northern Provinces of China* (1847), *A Journey to the Tea Countries of China* (1852), *A Residence Among the Chinese* (1857) and *Yedo and Peking* (1863).

He died in London in 1880, and is buried in Brompton Cemetery.

Numerous plants were named after Robert Fortune, including *Cephalotaxus fortunei*, *Cyrtomium fortunei*, *Euonymus fortunei*, *Hosta fortunei*, *Keteleeria fortunei*, *Mahonia fortunei*, *Osmanthus fortunei*, *Pleioblastus variegatus* 'Fortunei', *Rhododendron fortunei*, *Rosa* × *fortuneana* and *Trachycarpus fortunei*.

Roots

In any plant with a vascular (water-transport) system, roots are of vital importance. Not only do they anchor the plant safely in place and provide support, they also absorb water and essential plant nutrients from the soil or growing medium.

The actual absorption of water and mineral nutrients in solution is carried out by the many tiny hairs on the roots. Many root hairs form mycorrhizal relationships with soil fungi, which can be mutually beneficial to both partners. Some bacteria also associate with plant roots, namely those that can turn atmospheric nitrogen into a form that plants can use. Plants can benefit greatly from such relationships, sourcing vital nutrients with the help of a partner (see p. 58).

For a gardener, it is important to look after the roots of your plants, by ensuring the soil is in good condition and well prepared. Gardeners that do this will notice that their plants grow and establish quickly, as the roots are also growing strongly.

Root structure

A root system consists of a primary root and secondary (or lateral) roots. The primary root is not dominant, so the whole root system is fibrous in nature and branches in all directions to produce an extensive rooting system. This allows it to provide excellent anchorage and support to the whole plant, and to search out water and nutrients over a large area.

The main roots may become woody, and above a thickness of 2 mm (1/16 in) they generally lose the ability to absorb water and nutrients. Their main function instead is to provide anchorage and a structure to connect the finer fibrous roots to the rest of the plant.

Care must be taken, therefore, when transplanting any plant not to cause too much disturbance to the finer roots, as this can seriously impede recovery. Too much damage is likely to result in death.

The root systems of plants are just as varied as their top growth, although as this is rarely seen by the gardener it is seldom appreciated. Some may form an enlarged taproot that grows directly downwards, while others make a fine network of surface roots, as found in species of *Rhododendron*. The deepest roots are usually found in the plants of deserts and temperate coniferous forests, and the shallowest in tundra and temperate grasslands.

Desert plants show various strategies to cope with their extreme environment; those known as 'drought endurers' either have water stored away in succulent tissues, or they have massive root systems that cover a wide area, collecting as much of the scarce water as possible. The camelthorn bush (*Alhagi maurorum*) of western Asia has one of the most extensive root systems of any desert plant.

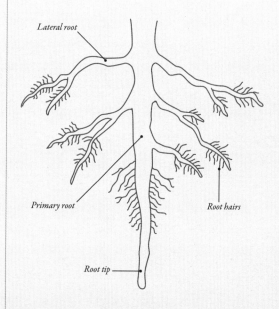

GROWTH, FORM AND FUNCTION

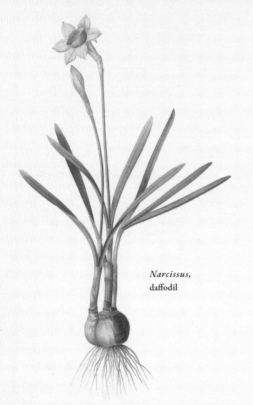

Narcissus,
daffodil

Aerial roots

Aerial roots are a common type of adventitious root, and are a common feature of tropical plants such as epiphytic orchids.

Aerial roots are used by some climbing plants, helping them to adhere to the climbing surface, be it a wall or another plant. Ivy (*Hedera*) and climbing hydrangeas are typical examples, and they can adhere so strongly that whole pieces of brickwork can be pulled away when overgrown plants are removed from old walls.

The seeds of strangler figs (*Ficus* spp.) germinate in the branches of other trees. They send out aerial roots that grow down towards the ground, and over time these become so numerous that the plant gradually envelopes and 'strangles' the host tree. Other plants send out 'prop roots' from stems to help provide support. In the garden, this can be observed on mature sweetcorn plants (*Zea mays*).

Roots often show special adaptations so that they can serve other functions besides water and mineral uptake and support, such as storing water and food reserves. Many plants can also die back to underground structures and enter into a period of dormancy. Two examples are the potato tuber and the daffodil bulb.

Adventitious roots

Like adventitious buds (see p. 52), these arise from unusual places, such as stems, branches, leaves or old woody roots. They can be important to gardeners when propagating plants from stem, root or leaf cuttings, as the aim is to encourage these severed plant parts to produce a new root system.

Hedera helix,
ivy

Ivies produce aerial roots, which are used to adhere the plant to the structure it is climbing or growing over.

Contractile roots

Contractile roots pull bulbs or corms, such as hyacinths (*Hyacinthus*) and lilies (*Lilium*), deeper into the soil by expanding and contracting. This serves to anchor the plant and keep them buried. The taproots of dandelions (*Taraxacum officinale*) serve a similar function.

Haustorial roots

Haustorial roots are produced by parasitic plants, such as mistletoe (*Viscum album*) and dodder (*Cuscuta*). They can absorb water and nutrients by penetrating the tissues of another plant.

Knee roots

'Knee roots' or 'pneumatophores' grow up from the ground into the air. They have breathing pores (lenticels) for the exchange of gases and are typically seen in swamp or waterlogged areas; the breathing pores enable the roots to survive underwater. In the garden, pneumatophores are most commonly seen on the bald or swamp cypress (*Taxodium distichum*), which is sometimes planted on the edge of large ponds.

Tuberous roots

Tuberous roots occur when a portion of a root swells for food or water storage, such as in sweet potato (*Ipomoea batatas*). They are distinct from taproots.

Taraxacum officinale, dandelion

Root relationships: root nodules and mycorrhizae

Nodules on roots can be clearly seen on most leguminous plants, i.e. those in the pea family (*Papilionaceae*). These nodules are specialised root structures that house bacteria from the *Rhizobiaceae* family as part of a complex symbiotic arrangement. The bacteria fix atmospheric nitrogen into a form that the plant can use, which reduces the plant's dependence on nitrogen in the soil. This makes legumes popular agricultural plants, as their reliance on the farmer for nitrogen fertiliser is much lower.

The two most important genera of bacteria within the *Rhizobiaceae* family are *Rhizobium* (generally seen on tropical and subtropical legumes, such as peanuts and soybeans) and *Bradyrhizobium* (mostly seen on temperate legumes, such as peas and clover).

The bacteria detect flavonoid chemicals released by the roots, and in turn release chemical signals of their own. The root hairs subsequently detect the presence of the bacteria and begin to curl around them. The bacteria then send infection threads into the cell walls of the root hairs and a nodule begins to grow, eventually forming an enlargement on the side of the root.

Cyanobacteria also associate with some plant roots, although the only flowering plants affected are from the genus *Gunnera* (such as the giant rhubarb, *G. manicata*). One of the most important cyanobacterial relationships must be that with the tiny water fern *Azolla* which is traditionally allowed to thrive in flooded rice paddies, where it acts as a biological fertiliser for the growing rice.

Mycorrhizal associations are fungal relationships. The term simply means 'fungus root', and it is estimated that over three-quarters of all plants have fungal partners. Fungi often form extensive mats through the soil with their filamentous hyphae, and by allowing the fungi to penetrate their roots, plants draw benefit from a pre-existing nutrient web. In return, the fungi also takes food from the plant.

Almost all orchids are known to form mycorrhizal relationships for at least part of their life cycle. Truffles are well known mycorrhizal fungi, and they are usually associated with certain kinds of trees – their preferred hosts. Gardeners can even buy trees already inoculated with truffle mycorrhiza, and dried mycorrhizal products are available to buy, which can be sprinkled into the root zone when planting any new plant.

Roots and the gardener

Roots respond to their environment and adjust their growth accordingly. As a rule, they grow in any direction where the correct environment of air, mineral nutrients and moisture exists to meet the needs of the plant. Conversely, they will stay away from dry, overly wet or other poor soil conditions, and areas of excessively high minerals, which can damage the sensitive root hairs. This is why excessive overfeeding can adversely affect root growth, and it is important to apply fertilisers carefully, only using the recommended quantities. A 'little bit extra for luck' often leads to problems.

As soil nutrients must be in solution for roots to be able to absorb them, it is important that the soil is kept moist. In dry soil, nutrients become much less available. Soil pH can also affect the availability of certain nutrients (see p. 144).

Damage to the soil (such as compaction) or poor drainage can have an adverse effect on root growth. Waterlogged soils can also deter rooting, and extended flooding may result in the death of the roots of plants that are not adapted to growing in such conditions. If the damage occurs in winter, the effects may not be seen until the next growing season when the plant

BOTANY IN ACTION

The roots of most garden plants are found relatively close to the soil surface where aeration and nutrient levels are more favourable for growth. Shallow-rooted plants, including shrubs such as *Calluna*, *Camellia*, *Erica*, *Hydrangea* and *Rhododendron*, are some of the first to suffer in dry soil, and they must be carefully cared for in such conditions. A layer of organic mulch over the soil can help, but if laid too thick it can 'suffocate' the roots. In their natural environment this is supplied in the form of leaf litter.

Camellia, camellia

attempts to put on new growth. Such a plant may be unable to draw water from the soil for the growth to be sustained, and death soon follows. It is important for this reason that any container to be used to grow a plant contains adequate drainage holes.

For more information on soils and soil compaction see chapter 6.

Prospero Alpini
1553–1617

It is thanks to Prospero Alpini that Europeans first heard about two of their daily staples – coffee and bananas – as he is credited with their introduction to Europe.

Alpini, sometimes referred to as Prospero Alpino among other spellings, was a botanist and physician, born in Marostica in the province of Vicenza, northern Italy.

He studied medicine at the University of Padua and after two years working as a physician in Campo San Pietro, a small town near Padua, he was appointed as medical adviser to Giorgio Emo, the Venetian Consul in Cairo, Egypt. This gave him the perfect opportunity to fulfil one of his main desires – to study botany, under more favourable conditions than he could find in Italy. As a physician, he was extremely interested in the pharmacological properties of plants.

Prospero Alpini is reputed to have been the first person to artificially fertilise date palms. He worked out sexual differences in plants that were adopted in the Linnaean taxonomic system.

He spent three years in Egypt, where he made an extensive study of the Egyptian and Mediterranean flora. He also had a business managing date palms and is reputed to have been the first person to artificially fertilise date palms. It was during this practice that he worked out sexual differences in plants, which was later adopted as the foundation of the Linnaean taxonomic system. He commented: 'The female date trees or palms do not bear fruit unless the branches of the male and female plants are mixed together; or, as is generally done, unless the dust found in the male sheath or male flowers is sprinkled over the female flowers.'

On his return to Italy, he continued as a physician until 1593, when he was appointed Professor of Botany at the University of Padua and Director of its Botanical Garden, the first to be established in Europe. Here he cultivated a number of species of Oriental plants.

He wrote several medical and botanical works in Latin, the most important and best-known being *De Plantis Aegypti liber* (Book of Egyptian Plants), which was a pioneering study on the flora of Egypt that introduced exotic plants to European botanical

Musa acuminata, banana

Prospero Alpini is credited with introducing bananas to Europe, producing the first European botanical account of the plant.

Title page from Alpino's *De Plantis Aegypti liber* (Book of Egyptian Plants), 1592, and a hand-coloured illustration of the fruit of *Adansonia digitata* (baobab), featured in the book.

circles. His earlier work, *De Medicinia Aegyptiorum*, contains the first mention by a European writer of the coffee plant, coffee beans and the properties of coffee. He also produced the first European botanical accounts of banana, baobab and a genus of the ginger family (*Zingiberaceae*), which was later named *Alpinia* after him by Linnaeus. *De Plantis Exoticis* was published after his death, in 1629. It describes exotic plants recently introduced into cultivation and was one of the first books to be devoted entirely to such exotics. It concentrated on Mediterranean flora, especially that of Crete, many of whose plants were described for the first time.

He died where his career started, in Padua, and was succeeded as Professor of Botany by his son, Alpino Alpini.

The standard author abbreviation Alpino is used to indicate him as the author when citing botanical names of plants.

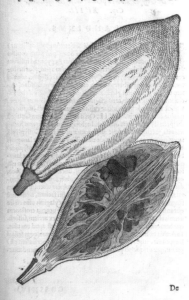

Coffea arabica, coffee

Stems

The structures that join the roots to the leaves, flowers and fruit are called stems, and they vary in thickness and strength. Internally, they contain all the vascular tissues, which distribute food, water and other resources through the plant (see pp. 96–97). Stems grow above ground level, although a corm (see p. 82) is technically a specialised underground stem.

The presence of water-conducting tissues has allowed vascular plants to evolve to a larger size than non-vascular plants (such as mosses), which lack these specialised conducting tissues and are therefore restricted to relatively small sizes.

Shoot
'Shoot' is a term used for new plant growth. As a shoot ages and thickens it turns into a stem.

Stalk
A 'stalk' is a name given to the stem supporting a leaf, flower or fruit. In botany, the correct term for a leaf stalk is a petiole. The stalk of a flower (or fruit) is called a pedicel. If the flower or fruit is held in a cluster then the stalk that supports the pedicels is called a peduncle.

Trunk
The trunk (or bole) is the main woody axis of a tree that supports the branches.

Modified stems

In some plants, stems show characteristic modifications, such as prickles and thorns to deter grazing and to help plants climb or clamber up other plants. Some stems are so specialised in form that they are unrecognisable as stems. They include cladodes and phylloclades, flattened stems that take on the appearance and function of leaves (as in cactus pads), and scapes – non-leafy stems that rise out of the ground to hold heads of flowers, as seen in lilies (*Lilium*), hostas (*Hosta*) and alliums (*Allium*). Pseudostems, as the term suggests, are not stems at all but stem-like structures made up of the rolled bases of leaves; the trunks of banana palms (*Musa*) are pseudostems.

Stems may be herbaceous or woody. There are no sclerenchyma cells in herbaceous stems, which means no woody growth (secondary thickening); they generally die back at the end of the growing season.

BOTANY IN ACTION

The function of stems

- Support and the elevation of the leaves, flowers and fruit above ground level, helping to keep the leaves in the light, the flowers close to pollinators and the fruit away from the soil, where it could rot.
- Transport of fluids around the plant via the vascular tissue.
- Storage of nutrients.
- The production of new living tissue from buds and shoots.

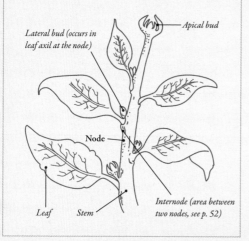

Lateral bud (occurs in leaf axil at the node)
Apical bud
Node
Leaf
Stem
Internode (area between two nodes, see p. 52)

Stem external structure

The typical anatomy of a stem includes the tip of the shoot – its apex – and the apical growth bud from which the stem grows and elongates. Below it, and attached to the stem, are the leaves, which are joined to the stem at an axil – the angle between the leaf and the stem. In each axil is an axillary bud, which produces sideshoots or flowers.

The position where the leaves and axillary buds are attached to the stem is called the leaf joint or node, and it is sometimes slightly swollen. The section of stem between two nodes is called the internode. It is typical to see either one or two leaves and/or buds attached at each node, sometimes three or more. Plants with just one bud per node are said to have alternating buds (since the buds usually alternate from left to right from node to node); plants with two or more buds per node are said to have opposite buds.

The way that buds are arranged on the stem can give clues to a plant's identity. For example, sweet gum trees (*Liquidambar*) and maple trees (*Acer*) are sometimes

Mentha, mint

Liquidambar styraciflua, sweet gum

Underground stems

These modified structures derive from stem tissue, but exist below the soil surface. They function either as a means of vegetative reproduction or as a food store, which is usually drawn upon during periods of dormancy brought on by cold or drought, for example. Being underground, the stems have some degree of protection.

A number of plant species also use underground stems to spread and colonise large areas, since the stems do not have to be supported and less energy and resources are needed to produce them as a result. Bamboos are a classic example, and gardeners need to be very wary when introducing these plants to the garden, making sure that they are 'clumpers' and not 'runners'. Even so, it is often a good idea to plant any bamboo within an area edged with a rootproof membrane. The same advice goes for mint (*Mentha*), which can run riot in a border.

For further information on the different types of underground stems, see pp. 82–83.

confused because of their similarly shaped leaves, but they are distinguished by their buds: alternating in *Liquidambar* and opposite in *Acer*. If it weren't for its opposite leaves, the Sydney red gum (*Angophora costata*) of eastern Australia could be easily confused with the many *Eucalyptus* trees of the same region.

Stem internal structure

Imagine a young or non-woody stem, cut across its width as if by a pair of secateurs. The cut surface will show a readily identifiable outer layer, called the epidermis, and within this – unseen to the naked eye – lies a ring of vascular tissue, made up of vascular bundles. At the pithy centre of the stem and around the vascular bundles lies a region of parenchyma tissue.

The epidermis covers the outside of the stem and usually functions to keep it waterproof and protected; some gas exchange may be able to take place through it to allow the cells within to respire and photosynthesise. The vascular tissue serves to transport water and nutrients around the plant, and as the cell walls are thickened it gives structural support to the stem.

Vascular bundles are made up two types of vessels: xylem and phloem vessels. The xylem is found on the inside layer of each vascular bundle (towards the centre of the stem) and is concerned with transport of water through the plant. The phloem is found on the outer

Magnolias are either grown as multi-stem shrubs or as single-stemmed trees.

layer of each vascular bundle and is used for transporting dissolved organic substances (such as nutrients and plant hormones). Sometimes when a stem is cut you can see droplets of moisture collect in a peripheral ring, which shows the position of the vascular tissue.

The main exception to this stem anatomy is found in monocotyledonous plants, where the vascular bundles are scattered throughout the stem rather than in a peripheral ring. Roots are also different to stems, with the vascular tissues arranged in the centre, like wire in a cable. Each vascular bundle is surrounded by a bundle sheath.

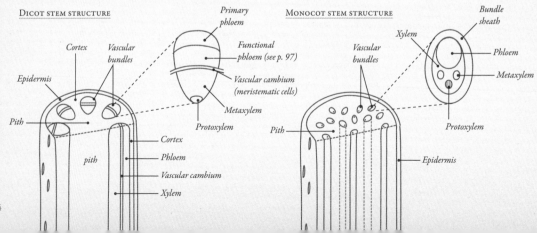

Getting woody with secondary growth

As stems age, the vascular cells divide laterally to cause radial growth, leading to an increase in circumference. Secondary xylem is produced to the inside, and secondary phloem to the outside. Secondary xylem cells produce wood, and the seasonal variations in growth seen in deciduous trees create the yearly tree rings.

The secondary phloem does not become woody and the cells remain alive. But between the phloem and epidermis a corky layer of cells begins to appear, forming a ring. A water-repellent substance called suberin is deposited in the cork cell walls, forming the bark and providing strength and reducing water loss. Lenticels are breaks in the cork layer, made up of loose cells; they allow the passage of gasses and moisture, and are clearly seen on the bark of many cherry trees (*Prunus*) as distinctive horizontal markings. Suberin is produced in abundance in the bark of the cork oak (*Quercus suber*), from where it derives its name.

Because of the scattered arrangement of the vascular bundles in monocotyledons, they grow in a different way. While radial growth is still possible, the larger monocots (like the palm trees) increase their trunk diameter by division and enlargement of the parenchyma cells, or through thickening meristems, derived from the apical meristem (growing tip). They either produce no secondary growth or, in the case of bamboos, palms, yuccas and cordylines, 'anomalous' secondary growth. If the dead wood of any of these plants is compared with that of any deciduous tree, for example, there is a huge difference: it is less dense and much more porous.

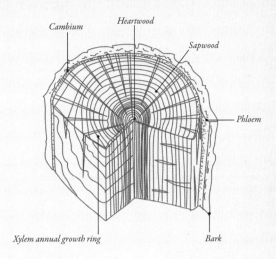

Prunus avium,
bird cherry

BOTANY IN ACTION

Girdling and ringing

Because phloem tubes sit on the outside of the xylem, and just under the bark, trees and other woody plants can be easily be killed by stripping away the bark in a ring on the trunk or main stem. This process is known as girdling or bark ringing.

Incomplete girdling (i.e., leaving about one-third of the bark intact) can be used to control a plant's growth. It can curb excessive leafy growth and help promote flowering and fruiting. It is a very useful process for unproductive fruit trees, with the exception of stone fruit. Mice, voles and rabbits can often girdle trees as they feed on its nutritious, sappy bark – where these animals are a problem, trees should be protected with some kind of netting or other physical barrier around the main stem.

Leaves

The ubiquitously abundant, thin and flattened green structure known as a leaf will be familiar to anyone. Gardeners often refer to 'foliage' when referring to the entire leafy mass of a plant, as it is the entire foliage effect that is most appreciated. The leaves of some plants, however, are appreciated more individually, such as those of *Hosta*.

Leaves are the 'power house' of the plant, as this is where most photosynthesis takes place – the chemical reaction whereby plants are able to produce the food they need to grow (see pp. 89–90). Actually, photosynthesis is able to occur in any plant tissue where there is green pigment, but the leaf is an organ specially adapted to this purpose.

The leaf must be well adapted to facilitate efficient photosynthesis if it is to do its job well. To this end, most leaves have a thin, flat shape – and therefore a large surface area – to maximise gas exchange and the amount of light they can absorb. Within the leaf, there are large air spaces to encourage easy gas exchange, and enveloping the leaf is a cuticle that must be transparent to allow the easy passage of light to the chloroplasts (where the photosynthetic reaction occurs) and yet waterproof so that the leaf does not dry out and wilt.

Some of the lower plants lack true leaves. Bryophytes (mosses and liverworts) and some other non-vascular plants produce flattened, leaf-like structures called phyllids, which are also rich in choroplasts.

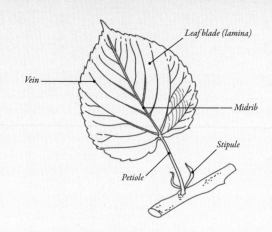

Leaf modifications

Anyone who has taken even the most cursory look at plants will notice that there is a huge diversity in leaf form. This is because the leaves of all plants show special adaptations to their natural habitat. The form a leaf takes often tells us a lot about the habitat from which the plant comes, and hence its cultivation requirements. Plant breeders sometimes create new leaf forms – different shapes, colours and textures – for ornamental effect.

The leaves of some plants are so specialised that they are almost unrecognisable by the standard definition and terms. Some are not even flat, such as in succulents where they are modified to store water, and some are found underground, such as bulb scales (which are used to store food). The harsh spines of cacti are modified leaves; they are not even photosynthetic, as this job has been taken up by the leaf-like, modified stems or cladodes (see p. 62). In carnivorous plants, the leaves take on a very specialised feeding function, as seen in the pitcher plants (*Nepenthes*) and Venus fly trap (*Dionaea muscipula*).

Dionaea muscipula,
Venus flytrap

GROWTH, FORM AND FUNCTION

The arrangement of leaves

Seen as a whole, the foliage of a plant is made up of many leaves, and although it may seem that there is no pattern to their arrangement, there often will be. Leaves are arranged so that the maximum light hits each one, and so that the leaves are not shading each other too much. For example, it is typical to see a spiral arrangement of leaves on plant stems to reduce the amount of shading, as well as hanging or pendent leaves as in willows (*Salix*) and *Eucalyptus*.

Different terms are used to describe how leaves are arranged on the stems, which is called phyllotaxis. Alternate leaves are attached singularly at each node, and they alternate direction. Opposite leaves are attached in pairs at each node, on opposing sides of the stem. Three or more leaves attached at the same point are usually referred to as being whorled. As with opposite leaves, successive whorls may be rotated by

Arum maculatum, wild arum

Salix × *smithiana*, silky-leafed osier

Some plants may even change their leaf shape as they grow and mature. *Eucalyptus* trees produce opposite pairs of rounded leaves when young, when their growth may be limited by the available light from surrounding plants. Their leaves change to an alternate arrangement of willow-like, pendent leaves when the plants reach a certain size, these being more suited to harsher light, higher temperature and drier conditions.

Further leaf modifications include bracts and spathes. Bracts are frequently associated with flowers, often brightly coloured to attract pollinating animals, performing an additional function to the petals, or sometimes replacing them. *Bougainvillea* and *Euphorbia pulcherrima* (poinsettia), for example, have large, colourful bracts that surround the smaller, less colourful flowers. Spathes form a sheath to enclose the small flowers of plants, as seen in palms and arums such as lords-and-ladies (*Arum maculatum*). In many arums the spathe is large and colourful, serving to attract pollinators to the small flowers, which are arranged on a thick stem called a spadix.

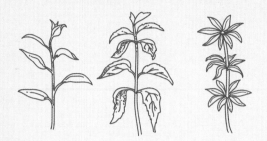

<u>Phyllotaxis (leaf arrangement)</u>

Alternate

Opposite

Whorled

The three commonest ways leaves are arranged on the stem are opposite (two leaves at each node), alternate (one leaf) and whorled (three or more).

half the angle between the leaves in the whorl in order to maximise the light hitting each leaf. Rosulate leaves are those that form a rosette.

Leaf external structure

The leaves of flowering plants have a typical structure that includes a petiole (leaf stalk), lamina (leaf blade) and stipules. Conifer leaves are typically needles, or tiny scales arranged on leafy 'fronds'. The leaves of ferns are called fronds.

Lamina

The lamina, or leaf blade, is the main part of the leaf. It can be described in two basic ways, depending on the way the lamina is divided: compound or simple. Compound leaves are divided into leaflets arranged along a main or secondary vein, or sometimes originating from a single point on the leaf stalk. Simple leaves may be deeply lobed or unusually shaped, but they are always entire, without leaflets. There are many different botanical terms used to described the shapes of leaves, with the most common listed as follows:

Simple leaves:

- ELLIPTIC – oval
- ENSIFORM – sword-shaped
- LANCEOLATE – lance-shaped
- LINEAR – long and narrow
- OBLONG – with parallel sides, two to four times longer than broad
- ORBICULAR – circular
- OVATE – egg-shaped
- PANDURIFORM – fiddle-shaped
- PELTATE – with the leaf stem joining at or near the centre on the underside
- RHOMBIC – diamond-shaped
- SAGITTATE – arrow-shaped
- SPATHULATE – spatula-shaped
- TRIANGULAR – with three distinct sides

Sagittaria sagittifolia, arrowhead

Trifolium pratense,
red clover

Compound leaves:

- PALMATE – in the form of a hand, as in horse chestnut (*Aesculus hippocastanum*)

- PINNATE – arranged along both sides of the mid-vein, like a feather, as in ash (*Fraxinus excelsior*); bipinnate leaves are twice divided in this way, such as in *Acacia*

- TRIFOLIATE – with just three leaflets, as in clover (*Trifolium*) and *Laburnum*

Petiole

The petiole is the stalk attaching the lamina to the stem; it usually has the same internal structure as the stem. Not all leaves have a petiole, which is typical of monocotyledons, and those missing one are said to be 'sessile' – or 'clasping', when they partly surround the stem. In rhubarb (*Rheum* × *hybridum*), the petiole is the edible part of the plant.

In some plants, such as in many species of *Acacia*, the petioles are flat and wide and are known as phyllodes. The true leaves may be reduced or absent altogether and the phyllode serves the purpose of the leaf. Phyllodes are often thick and leathery and help the plant survive dry environments.

The midrib is the main vein of a leaf, a continuation of the petiole. In pinnately veined leaves, the midrib is the central vein to which the leaflets are attached; in palmately veined leaves, it may be absent.

Stipules

Stipules are outgrowths seen on either side, or sometimes only one side, of the base of the petiole. They are commonly absent or inconspicuous, or reduced to hairs, spines or secreting glands.

Leaf internal structure

Only at the microscopic level are the true wonders of leaves seen. True, many may have wonderful shapes and patterns, but the internal workings and chemistry of leaves is nothing short of miraculous. Virtually all animal life on Earth depends on them, as they possess the ability of turning sunlight into food. No animal is capable of this.

On the underside of leaves, small pores called stomata (singular: stoma) are found. Note that they are sometimes also found on the upper leaf surfaces and other parts of plants, but it is on the leaf undersides that they are in their greatest density. It is through the stomata that oxygen, carbon dioxide and water vapour pass to and from the cells within the leaf. In essence, these are the pores through which a plant can 'breathe'.

Stomatal pores are open during the day, and at nightfall they close, when photosynthesis pauses. It is the action of two guard cells, which border each pore, that allows the stomata to open and close. Guard cells operate through increases and decreases in water pressure, which is controlled by the level of solutes

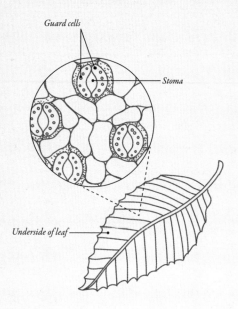

Evergreen and deciduous leaves

An evergreen plant is one that has leaves all year round. Evergreens include most species of conifers, 'ancient' gymnosperms such as cycads, and many flowering plants, particularly those from frost-free or tropical climates.

Deciduous plants lose all or nearly all of their leaves for part of the year. This leaf loss usually coincides with winter in temperate climates. In tropical, subtropical, and arid regions, deciduous plants may lose their leaves during the dry season or during other adverse conditions.

inside each guard cell, which is in turn influenced by the level of light. In darkness, the concentration of solutes falls and water moves out into the surrounding cells, forcing the guard cells to collapse and close.

Guard cells help to regulate the amount of water lost by the leaf. They close when the rate of photosynthesis is low, and they also close during periods of dry weather or drought. In plants that grow in areas with low or irregular supplies of water (xerophytes), the stomata are sunken into the leaf epidermis, which serves to trap moist air around the pore and therefore reduce evaporation.

Epidermis

The epidermis is the outer layer of cells covering the leaf, which separates the inner cells from the external environment. It provides several functions, but the main ones are protection against excessive water loss and the regulation of gaseous exchange. It is covered with a transparent waxy cuticle that helps prevent water loss and, as a result, is usually thicker on plants from dry climates. Most evergreen plants also have a thick cuticle, and this is often glossy to reflect the sun's heat and consequently reduce evaporation.

BOTANY IN ACTION

Evergreens

Evergreens do lose their leaves, but not all at the same time as deciduous plants do. Under any evergreen you will find plenty of leaf litter, and if left uncollected it will decompose and the nutrients returned to the soil. Some deciduous trees, including beech (*Fagus sylvatica*), hornbeam (*Carpinus betulus*) and a few species of oak (*Quercus*), retain their desiccated leaves through winter, only to drop them in spring as new growth begins. This is most often seen when they are clipped as hedges and it can be an ornamental feature in its own right.

Fagus sylvatica, common beech

Flowers

The flower is the reproductive structure of flowering plants (angiosperms), which ultimately gives rise to the seeds and fruit. There is more information on angiosperms on pp. 25–27 and sexual reproduction on pp. 110–115. Flowers are also referred to by gardeners as blooms or blossom, and there is a huge diversity in shape and form.

The flower provides a mechanism for pollen from the male part of a flower to fertilise the egg cells in the female part of the flower. The flowers may develop or be constructed in such a way as to encourage cross-pollination between different flowers, or permit self-pollination within the same flower (see chapter 4).

Many plants have evolved to produce large, colourful flowers that are attractive to animal pollinators, while others have evolved to produce dull, scentless and nectarless flowers that are pollinated by the wind. Two contrasting examples would be the foxglove (*Digitalis purpurea*) and the golden oat grass (*Stipa gigantea*) – one insect-pollinated, the other not, but both with perfectly effective strategies. For more information on pollination, see chapter 4.

Bellis perennis, common daisy

Digitalis purpurea, foxglove

The arrangement of flowers

A group or cluster of flowers is referred to as an inflorescence. Botanists recognise many different forms of inflorescence, yet most gardeners simply refer to any cluster of flowers as a 'flowerhead' – a usefully imprecise term.

There are as many types of inflorescence as there are types of flower, but the main types are as follows:

Capitulum
A tightly packed head of tiny flowers (florets). It may resemble a single flower, as in sunflowers (*Helianthus*) and daisies (*Bellis*).

Corymb
A fairly flat-topped flowerhead, where the individual flowers arise from different points on the stem, as in common hawthorn (*Crataegus monogyna*).

Echium vulgare,
viper's bugloss

Spadix

A spike with a fleshy stem with many tiny flowers, most commonly associated with a spathe – a brightly coloured modified leaf or bract, as in lords-and-ladies (*Arum maculatum*).

Spike

A flowerhead with a main axis, from which numerous unstalked flowers arise. Grasses typically produce their flowers on spikes.

Umbel

A flat-topped flowerhead, a bit like a corymb, but where the flower stalks all arise from the same point at the top of the main axis. Umbels may be simple or compound (as in the large herb *Angelica archangelica*).

Angelica archangelica,
angelica

Angelica has a compound umbel: its flowerhead is made up of several umbels.

Cyme

Each branch ends in a flower, and younger flowers are produced on a succession of sideshoots. In a monochasial cyme, which is usually in the shape of a spike, but may be one-sided or curled at the tip (as in some *Echium*), the lower flower opens first. In a dichasial cyme, which is often dome-shaped, the central flower opens first.

Panicle

A flowerhead with a main axis, from which many further branching stems arise. Such flowerheads can be fairly intricate, as in *Gypsophila*.

Raceme

A flowerhead with a main axis, from which individual flowers are borne on short stems, as in foxgloves (*Digitalis*).

Leaf-like bracts may also be a feature of some inflorescences. They sometimes form part of the flowerhead, as in daisy flowers (*Bellis perennis*) or they may be brightly coloured as in poinsettias (*Euphorbia pulcherrima*). In more complex flowerheads, such as compound cymes, the smaller bracts found on the branching sideshoots are known as bracteoles.

The structure of flowers

In chapter 1 (see p. 27), the basic structure of a flower was broken into four constituent parts: the sepals and petals (perianth), the stamen (male parts) and the pistil (female parts). These are arranged in whorls, with the sepals outermost and the pistil innermost.

The extensive differences in flower form are among the main features used by botanists to establish relationships between plant species. It is a general rule that more ancient plants such as buttercups (*Ranunculus*) have a greater number of flower parts than more highly derived plants like those of the mint family (*Lamiaceae*) or the orchid family (*Orchidaceae*) with their deceptively 'simple' flowers.

In the majority of plant species, individual flowers produce both functional male and female organs and are described as hermaphrodite or bisexual. Some species, however, and sometimes varieties within a species, have flowers that lack one or other reproductive organ and are known as unisexual. In plants with unisexual flowers, if these are carried by the same individual plant, the species is called monoecious. If they are on separate plants, the species is called dioecious. The monoecious type is much more common. *Skimmia* shrubs are dioecious, as are most hollies (*Ilex*); in gardens the female forms are usually more widely grown as they bear both flowers and fruit.

Animal-pollinated flowers often produce nectar, a liquid rich in sugars produced from glands called nectaries. These are usually located at the base of the

Epidendrum vitellinum,
yolk-yellow prosthechea

perianth, that pollinators attracted to the nectar brush the anthers and stigma on their approach, so ensuring pollen transfer on every visit. Pollinators attracted by nectar include bees, butterflies, moths, hummingbirds and bats.

In cultivation, plant breeders sometimes exploit natural mutations in plants, whereby some or all of the sexual flower parts are converted into extra petals. Depending on the extent of this mutation, we see 'double' or 'semi-double' flowers. Roses are a common example. Since double flowers contain few or no stamens, they cannot be expected to bear fruit.

Seeds

Seed development occurs directly after fertilisation (see p. 115). As discussed in chapter 1, angiosperms (flowering plants) produce enclosed seeds, protected within a carpel, and gymnosperms produce 'naked' seeds with no special structure to enclose them. Naked seeds are typically (but not always) carried uncovered on the bracts of the cones.

As the seeds of flowering plants mature, the carpel also begins to ripen into a hard or fleshy structure; as a whole, this unit becomes the fruit (see p. 78). The fruit may remain a combined entity right up until the moment that the seed(s) within it germinate, or a seed may separate from its housing. Fruit that opens to release its seeds is termed dehiscent, whereas fruit that does not open is indehiscent. The exact mechanism depends on the life strategy of the plant concerned, and it is largely to do with protection and dispersal of the seed (see pp. 78–80).

To gardeners, the word 'seed' can be extended to 'seed' potatoes (which are, in fact, stem tubers), and some of the things that gardeners sow are actually dry fruits (which contain the seeds), such as grass 'seed' or the corky fruits (sometimes called seed clusters) of beetroot (*Beta vulgaris*). Most seed for commercial production is carefully screened and 'cleaned' from non-seed debris, so that when you buy a packet of seeds that is essentially what you get.

The purpose of seeds

In evolutionary terms, seeds are an important innovation that has led to the huge success of angiosperms and gymnosperms (collectively known as the spermatophytes – the seed-bearing plants). One main advantage they have over the spores of lower plants (such as ferns and mosses) is that they are, as a rule, far more durable, being able to endure extended periods of dormancy and harsh environments.

Although they are not the only means by which plants reproduce themselves, seeds are often able to disperse themselves over great distances (take, for example, the tufted airborne seeds of dandelions (*Taraxacum*) and enable a plant to colonise new, faraway locations. Seeds are also the product of sexual reproduction (with a few exceptions), and thus provide genetic variability, which not only benefits natural populations but also the work of plant breeders.

Annual plants use seed production as a form of dormancy. The seed will only germinate when conditions are again favourable, and the subsequent plant will grow, flower and set seed, thereby completing another cycle. Seed banks around the world collect seed and keep it in a dormant state, usually at very low temperatures. Some seed banks are concerned with the conservation of agricultural seed diversity (such the Svalbard Global Seed Vault on the island of Spitsbergen in Norway), and others with wild species (such as the Millennium Seed Bank Project in West Sussex, England). Such efforts could ensure the conservation of entire species or valuable crop plants in the event of a global catastrophe.

Seed structure

Any angiosperm seed consists of two essentials: an embryo and a seed coat (testa). A food store (endosperm) could also be included as an essential component, were it not for the fact that some highly specialised seeds have overcome the need for an endosperm (see next page).

The embryo consists of a plumule (the nascent stem), a radicle (first root), and either one or two seed leaves (depending on whether it is the seed of a monocot or a dicot).

GROWTH, FORM AND FUNCTION

The testa is the seed covering, enclosing its contents but perforated at one point called the micropyle. Its main function is to protect the embryo from physical damage and dessication. It can be thin and papery, as in peanuts (*Arachis hypogaea*), or very tough, as in the coconut (*Cocos nucifera*). The micropyle allows passage of oxygen and water upon germination. A scar, known as a hilum, is often present where the seed was once attached to the ovary wall.

Some seed coats have extra features such as hairs (as in cotton, *Gossypium*), arils (such as the fleshy substance attached to individual pomegranate seeds, *Punicum granatum*), or fatty attachments known as elaiosomes. These often aid dispersal of the seed.

The endosperm is a mass of food-storing tissue, and its purpose is to supply the seedling that grows from the embryo as it germinates, as well as providing a source of energy during dormancy. The seeds of orchids (family Orchidaceae) have no endosperm and will only germinate in the presence of a suitable fungal counterpart, which forms a close association with the developing seed and provides its nutriment.

Orchid seeds must represent an extreme of seed evolution; they have reduced their contents to the bare minimum so that their seeds are almost dust-like, with each plant producing an uncountable number each season. They are dispersed in the wind. The tiny, pin-prick seeds of vanilla (*Vanilla planifolia*) are an example.

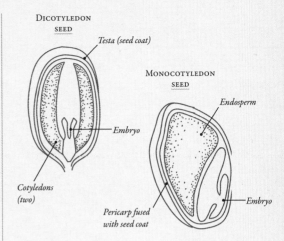

Punica granatum, pomegranate

BOTANY IN ACTION

Dispersal of seeds

Plants that produce small seeds can produce many seeds, which is one strategy for ensuring that at least one of them will land in a favourable place. Those with large seeds produce fewer seeds, investing more resources and energy in each one. Their dispersal strategies are usually far more specific. The largest seed of all is the coco-de-mer (*Lodoicea maldivica*), which can weigh up to 30 kg (66 lb).

Small seeds ripen faster and often spread further. Larger seeds may result in larger and stronger seedlings that can out-compete other plants. The strategies that plants adopt are huge and varied; it is impossible to say that one is better than another.

Richard Spruce
1817–1893

Richard Spruce was a great Victorian botanical explorer, spending 15 years exploring the Amazon River, from the Andes to its mouth. He was one of the first Europeans to visit many of the places along the river.

Spruce was born near Castle Howard in Yorkshire, England. From an early age, he developed a great love of nature and natural history, and making lists of plants was among his favourite pasttimes. At the age of 16 he drew up a list of all the plants he had found in the area where he lived. These were arranged alphabetically and contained 403 species, a task that must have taken some time and obviously been a labour of love for one so young. Three years later he had drawn up the *List of the Flora of the Malton District*, containing 485 species of flowering plants, many of which are mentioned in Henry Baines' *Flora of Yorkshire* (1840).

He became particularly interested in bryophytes – mosses and liverworts – and was an acknowledged expert, with a sizeable herbarium of his own comprising specimens from the British Isles and further afield.

This early interest in plants led him to carry out a major expedition to the Pyrenees in 1845 and 1846. His intention was to finance his expedition by selling sets of flowering plant specimens, but the little-known bryophytes did not create much interest. During his collections from this region he discovered at least 17 species new to science, and increased the bryophyte list of the area from 169 species to 478.

Two years later, he was approached by William Hooker, Director of the Royal Botanic Gardens, Kew, to carry out the botanical exploration of the Amazon on Kew's behalf. Despite failing health, he agreed, as it was

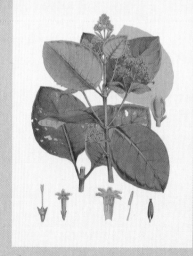

Richard Spruce amassed a huge collection of plants and other objects from his travels.

Cinchona pubescens, quinine

Richard Spruce's collection of quinine bark helped millions fight against malaria.

'The largest river in the world runs through the largest forest. Fancy — two millions of square miles of forest, uninterrupted save by the streams that traverse it.'
Richard Spruce

a huge opportunity for him. Again, he financed the trip by selling sets of specimens to interested naturalists and institutions in Europe.

During the following 15 years his travels took him along the Amazon River to Brazil, Venezuela, Peru and Ecuador, where he collected more than 3,000 specimens, making him a major contributor to the knowledge of the flora of the area. Spruce was also a keen anthropologist and linguist, learning 21 different languages while he was there, and collected many locally produced items of ethnobotanical, economic and medical interest as well as plants.

He discovered *Banisteriopsis caapi* and observed its use among the Tukanoan Indians of Brazil. This is one of the two ingredients in ayahuasca, a brew of psychoactive ingredients used by the shamans of the indigenous western Amazonian tribes in religious and healing ceremonies.

Of the thousands of plants that Spruce collected, the most important were undoubtedly *Cinchona*, or *Quina*, from Ecuador, a genus in the family *Rubiaceae* from which quinine bark was harvested. Native South Americans used the bark as treatment for malaria. Spruce provided seeds of the trees to the British government, making bitter bark quinine widely available for the first time, and so enabling the establishment of plantations in British colonies across the world and helping millions fight against malaria. After returning to England, he wrote *The Hepaticae of the Amazon and the Andes of Peru and Ecuador*.

Other published work of his includes descriptions of 23 new British mosses, about half of which he had discovered, in the *London Journal of Botany*, and his 'List of the Musci and Hepaticae of Yorkshire' in *The Phytologist*; in this he recorded 48 mosses new to the English Flora and 33 others new to that of Yorkshire.

Spruce received a doctorate from the Academia Germanica Naturae Curiosorum (German Academy for those who are Curious about Nature) in 1864, and later became a Fellow of the Royal Geographical Society.

Richard Spruce was particularly interested in bryophytes – mosses and liverworts – from an early age and became an acknowledged expert.

The plants and other objects collected by Spruce form an important botanical, historical and ethnological resource. The Richard Spruce project is a joint initiative between Royal Botanical Gardens, Kew and The Natural History Museum, comprising specimen location and databasing, specimen imaging and transcription and imaging of Spruce's original notebooks. More than 6,000 specimens have been completely databased and imaged, making information available to botanists, historians and others interested in the exploration of the Amazon and Andes.

Sprucea (now *Simira*, in the family *Rubiaceae*), and the liverwort *Sprucella* are named after him. The standard author abbreviation Spruce is used to indicate him as the author when citing botanical names of plants.

Fruit

To any gardener, 'fruit' usually refers to the sweet and fleshy crops borne on trees and shrubs through the summer and autumn. These would include top fruit such as quinces, soft fruit such as raspberries, and bush fruit such as redcurrants. They may also include the ornamental fruit of crab apples (*Malus*) or dogwoods (*Cornus*). Further to this are the fruiting vegetables, such as pumpkins (*Cucurbita*) and peppers (*Capsicum*).

To the botanist, however, all flowering plants are capable of producing fruit. It is a strict term used to define the structure that matures from a flower upon fertilisation of its ovary. A fruit contains the seed(s), and these are enclosed by a pericarp, a fleshy or hard coating that serves to protect the seed and aid its dispersal. The pericarp forms from the ovary wall.

Rubus idaeus, raspberry

The fruit of chilli and sweet, or bell, peppers (*Capsicum annuum*) are available in a wide range of colours, shapes and sizes.

How the fruit is used to disperse the seeds it contains

Anyone reading this book who has ever eaten a fruit will have been an unwitting agent in the dispersal of its seeds. It could be said that many popular fruits – apples, tomatoes, raspberries to name but a few – owe their success to the fact that their fruit is so delicious (and nutritious) to eat.

Animals

Fruit passing through the gut of an animal soon becomes separated from its seed, and will be deposited – most likely – some distance away from its parent plant in a heap of ready-prepared compost. In the process, the seed will have been washed and cleaned, and removed of any chemical inhibitors that prevent the seed from germinating while it is still in the fruit.

This is only one in a great number of seed dispersal mechanisms, however, and the great variety in fruit structure is testament to this. Some fruit, such as burdock (*Arctium*) and the New Zealand burr (*Acaena*) are covered with spikes or hooked burrs, which stick to the hairs or feathers of passing animals and can therefore be carried for miles and miles. The round red fruit of the Italian buckthorn (*Rhamnus alaternus*) is eaten by animals in its native habitat, but this is only the first stage in the dispersal of its seeds. Once they have passed through the animal's gut, the exposed seeds crack open in the hot sun to expose an oily, edible covering called an elaiosome. Ants find this very attractive, and they collect the seeds and carry them underground, where they will germinate the following year.

Dispersal by air

Animals are not the only seed dispersers; seeds are also scattered by the elements. Fruits carried in the air are either tiny or elongated and flattened out to make them easier for the wind to carry them long distances. Others develop wings or blades, such as the 'helicopters' of maple trees (*Acer*) or the 'parachutes' of dandelions (*Taraxacum*).

Dispersal by water

On the water, buoyant coconuts and mangrove seeds can float thousands of miles in the oceans. So successful is the coconut (*Cocos nucifera*), with its huge seeds lined with rich food (coconut meat) and partially filled with water, that it has managed to colonise almost all the beaches in the tropics. The seeds of the sea bean (*Entada gigas*) are sometimes seen washed up on European beaches, far away from their native home in the Caribbean and other tropical locations. They remain viable for up to a year.

Fire

In habitats where fire is a regular feature, we see seed pods that resolutely refuse to release their seed unless extremely high temperatures are reached. In the eucalyptus forests of southern Australia, seedlings find it hard to get a foothold because of the thick and unusually tall ground cover, where tree ferns can stand up to 3 m (10 ft) tall. Even when a tree falls, it may not be enough to create a break in this ferny canopy. When a bush fire burns through the forest, however, which is a natural feature of this continent, the eucalyptus seed pods burst open, and the seeds scatter all over the scorched earth. These seeds may have been waiting for years, but within a week they will have germinated and the long process of regeneration begins.

Cocos nucifera,
coconut

The buoyant coconut fruit can float for thousands of miles, so successfully dispersing the seeds over a very wide area.

Dispersal by gravity

Some fruit are large and heavy enough to roll down hillsides, such the heavy and thick fruit capsules of the brazil nut tree (*Bertholletia excelsa*). The fruit is the size of a coconut and the seeds within (brazil nuts) are packed in segments. When ripe, the capsule falls from the trees with a heavy thud, which may be sufficient to crack open its weak 'lid'. If the capsule falls on a slope, it will probably roll some distance away from its parent, the seeds inside sometimes spilling out as it rolls. The capsules are sometimes known as 'monkey pots', because the native capuchin monkeys will try to extract the nuts through the small lid; the monkeys will often carry the capsules around with them while they figure out how to extract the nuts. Native rodents often gnaw away at the capsule to get to the seeds, which they will store around the forest as squirrels do with nuts. Inevitably, some get forgotten and they will eventually germinate if there is a break in the forest canopy caused by a fallen tree.

Impatiens glandulifera,
Himalayan balsam

Bertholletia excelsa,
brazil nut

Explosive dispersal

Passing contact, a build-up of pressure caused by the fruit drying out, or a combination of both can trigger fruits to literally explode, catapulting or flinging their seeds into the air. This is seen in the Himalayan balsam (*Impatiens glandulifera*), which has become a rapid coloniser outside its native range by virtue of its explosive fruit. So successful is it that it is now listed as an invasive weed in the British Isles according to Schedule 9 of the Wildlife and Countryside Act in England and Wales, making it an offence to plant or sow this species in the wild. Gardeners would be wise also to avoid planting it in their gardens. Explosive fruit is also seen in the squirting cucumber (*Ecballium elaterium*), which is an astonishing plant common in Mediterranean scrub, and in several other common garden plants such as witch hazels (*Hamamelis*), brooms (*Cytisus*) and geraniums, although their mechanisms might not be so readily observed.

The many different types of fruit

Any casual observer of plants or their fruit, or even fruit fanatics, will be staggered by the enormous variety of fruit forms that nature has to offer. Botanists classify them into three main groups: simple, aggregate and multiple.

Simple fruit can be dry or fleshy, and results from the ripening of a single carpel or multiple carpels united in a single ovary. Simple dry fruit include achenes (containing a single seed, as in the globe artichoke, *Cynara cardunculus*), samaras (winged achenes, as in maples, *Acer*), capsules (formed from two or more carpels, as in love-in-a-mist, *Nigella*), grains (as in wheat, *Triticum*), legumes (commonly called pods, like those of peas, *Pisum sativum*), nuts, and siliquae (the multi-seeded pods of plants in the cabbage family, *Brassicaceae*).

Fleshy simple fruits include berries (where the entire ovary wall develops into a fleshy pericarp, as in blackcurrants, *Ribes nigrum*) and drupes (where the inner part of the ovary wall develops into a hard shell – the pit or stone – and the outer part into a fleshy layer, as in members of the cherry genus, *Prunus*, and olives, *Olea europaea*).

A single flower may comprise a number of separate carpels that fuse together as they grow to form a larger unit. Separately they are known as fruitlets, together they become an aggregate fruit. Achenes, follicles, drupes and berries are capable of forming aggregate fruit. Blackberries and raspberries (*Rubus*) are aggregate fruit consisting of drupelets.

Strawberries (*Fragaria × ananassa*) are a type of aggregate fruit, except that the fruit are not joined together by the fusion of the carpels; they are instead joined by another part of the flower (the receptacle), which has enlarged and become fleshy as if it were part of the fruit. Pome fruit (apples, pears and quinces) are another example. Note that not all flowers with multiple carpels necessarily go on to form aggregate fruit; they may remain separate as in the burred achenes of the common weed herb bennet (*Geum urbanum*).

Multiple fruits are formed from a flowerhead or inflorescence. Each flower produces a fruit, which then fuse into a larger, single fruit. Fleshy examples include pineapples (*Ananas comosus*) and mulberries (*Morus*). A common example of a dry multiple fruit is the spiky 'balls' produced by plane trees (*Platanus*). Gardeners might often wonder about fruit without seeds: bananas, for example, and seedless grapes. For those curious enough to ask the question, the answer lies in parthenocarpy, meaning 'virgin fruit'. It sometimes occurs as a mutation and results in the formation of fruit without fertilisation. Commercially, this is exploited to produce seedless oranges, bananas, aubergines and pineapples. Seedless grapes are not technically parthenocarpic; in this instance normal fertilisation takes place but the embryos are aborted soon after, leaving the remains of undeveloped seeds. This is known as stenospermocarpy.

Ananas comosus,
pineapple

Bulbs and other underground food storage organs

A number of perennial plants produce specialised food storage organs that ensure they can live for many years. They often die back to these underground structures during periods of dormancy. This allows the plants to survive periods of adverse environmental conditions, such as cold winters or dry summers. Plants also use these organs as a way to spread and multiply.

Many storage organs are modified stems and so bear similarities with them, such as an apical growing point, buds and modified leaves (sometimes known as scales). The eyes of potato tubers are a type of bud.

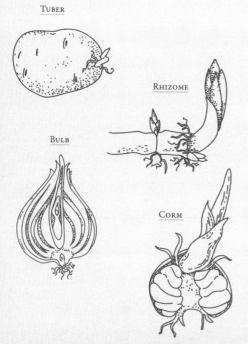

TUBER

RHIZOME

BULB

CORM

Bulbs

A true bulb is essentially a very short stem with a growing point enclosed by thick, fleshy, modified leaves called scale leaves. These store the food reserves to sustain the bulb through dormancy and subsequent re-emergence. In most bulbs the scales are thin and closely packed together, as in daffodils (*Narcissus*), but in others like lilies (*Lilium*), they are loose and swollen. The foliage and flower buds grow up from the centre of the bulb; the roots grow from the bulb's underside.

When planting bulbs it is useful to recognise which end is the top of the bulb, as this needs to be planted uppermost. It can sometimes be hard to tell; bulbs planted the 'wrong way up' usually right themselves, but it can put an extra strain on the bulb's energy reserves. As a rule, bulbs are best planted at three times their own depth. Summer and autumn bulbs should be planted in the spring, and spring bulbs in the autumn.

Corms

A corm is the swollen, underground base of a solid stem, seen in crocosmias and gladioli. It stores food reserves and is surrounded by scale leaves, which protect the corm. There is at least one bud at the apex of the corm, which will develop into the leaves and the flowering shoot.

Bulbs and corms are often confused as they look very similar. One of the major differences is that bulbs are made up of many fleshy scales; corms are solid structures (essentially packed with parenchyma tissue). Corms also tend to have much shorter lifespans, being replaced by a new corm that forms on top of the old one. Many tiny cormlets also form around the base of the corm, which form multiple new stems.

Tubers

Tubers are the swollen, enlarged, fleshy ends of an underground stem. They have 'eyes' comprising a cluster of buds and a leaf scar; together these are the equivalent of the nodes of a typical stem. They can arise anywhere on the tuber's surface, but are usually most densely arranged at one end – at the end opposite to where the tuber was attached to the parent plant. When chitting (sprouting) potatoes in spring, in preparation for planting, it is the end with the most buds on that should be pointed upwards. It follows that they are consequently planted this way up in the soil. The tuber usually shrivels once a new plant has grown from it, and eventually new tubers will grow from the new plant.

As well as the potato, other common garden tubers include tuberous begonias and cyclamens. Sweet potatoes and dahlias also grow tubers, although these are technically root tubers, unlike the above examples, which are stem tubers. Root tubers are essentially swollen roots; there are no nodes or eyes, which are structures derived from stems. Instead, adventitious buds form from either end, from which the roots and shoots grow. Some day lilies (*Hemerocallis*) form root tubers.

Rhizomes

Rhizomes are stems that grow horizontally on or just below the soil surface. The stem appears segmented as it is made up of nodes and internodes, from which the leaves, shoots, roots and flower buds develop. The main growing point is at the tip of the rhizome, but other growing points also appear along its length so, like a tuber, several shoots can appear at once.

Ginger (*Zingiber officinalis*) is a rhizome, as are the thickened, ground-covering stems of bearded irises. Many plants spread rapidly by underground rhizomes, such as those of the shuttlecock fern (*Matteuccia struthiopteris*) and the vigorous bamboo *Sasaella ramosa*.

Gardeners can separate rhizomes into pieces, and as long as each piece has a growing point, it can be encouraged to establish into a new plant. Shorten any leaf blades roughly by half and discard any dead material, then replant the rhizomes at the same depth that they were originally growing at. It may take a few seasons for the new plant to fully establish itself.

Stolons

Stolons are similar to rhizomes, in that they are also horizontal stems that run at, or just below, the soil surface with rooting and shooting nodes. They differ from rhizomes as they are not the main stem of the plant – they arise from a main stem and new plants are produced at the ends. When produced above ground level they are called runners.

BOTANY IN ACTION

Runners and stolons

Strawberries are famous for producing runners, and these are easily grown on. Once rooted, they can be separated from the parent plant and replanted. Many weeds spread themselves rapidly by runners and stolons, such as the creeping buttercup (*Ranunculus repens*).

Ranunculus repens, creeping buttercup

Cosmos bipinnatus, cosmos, Mexican aster

Chapter 3

Inner Workings

Plants grow. Like all living things, they are made up of cells, and these cells divide and expand. The nutrients needed to make this possible come from the soil, and the energy required to construct new cells primarily comes from sunlight.

In its most basic form, a plant is like a drinking straw, drawing water and nutrients out of the soil and up through the stems to the leaves, where the water is lost through evaporation (properly termed 'transpiration'). Nutrients circulate around the plant through these vascular tissues, as do the hormones that regulate plant growth.

For most gardeners, this simple overview is probably all they need to know about the inner workings of plants. Yet what goes on inside the cells is intricate and fascinating, and our knowledge of it is the result of centuries of inquiry and discovery by scientists.

Cells and cell division

Modern cell theory states that all living matter is composed of cells, all cells arise from other cells, all the metabolic reactions of an organism occur within cells, and that each and every cell (with a few exceptions) contain all the hereditary information needed to make a new plant.

The cell wall

Many plant cells are surrounded by a cell wall. Young, actively growing plant cells will only have a thin primary cell wall; many mature plant cells, especially those in xylem tissues that have finished growing, will lay down a secondary cell wall.

Gentiana acaulis, stemless gentian

Primary cell wall

The primary cell wall plays an important role in a plant's strength and support. Fully hydrated (turgid) cells exert an outward pressure on the wall, causing the cells to swell, but they are prevented from bursting by the presence of cellulose, which is both strong and elastic. The pressure of turgor keeps the stems upright, and when a plant starts to dry out, turgor pressure is reduced and it starts to wilt.

Compared with the rest of the cell, the primary cell wall is actually quite thin, just a few micrometres thick. Up to one-quarter of a cell wall is made up of long cellulose fibres, and because of their parallel arrangement they have as much tensile strength for their weight as steel wire. The cellulose fibres are embedded in a matrix of other materials, including hemicelluloses and polysaccharides.

The primary cell wall is well adapted to growth, as the cellulose fibres are able to move within the matrix in response to expansion. The cell wall stretches as the cell grows or swells, and new material is added to the wall so that it maintains its thickness. It is freely permeable to water and dissolved nutrients.

Secondary cell wall

In many plants, after a cell has stopped enlarging, it begins to lay down a secondary cell wall. In mature xylem tissues, the secondary cell wall provides support for the plant, and, in all but a small portion of cells in wood and cork, once the secondary wall is in place the cell within dies so that only the wall remains.

Secondary cell walls are much thicker than primary cell walls and they consist of about 45% cellulose, 30% hemicellulose, and 25% lignin. Lignin is not easily compressed and resists changes in form. In other words, it is a lot less flexible than cellulose.

The combination of lignin and cellulose fibres in the secondary wall is similar to steel rods embedded in

Cell Structure

Viewed under a microscope, a plant cell contains the following six distinctive structures:

1. Cell wall

The cell wall is a thick and rigid structure containing fibres of cellulose. It firmly fixes the position and shape of the cell, gives the cell protection and provides support. On its inner surface lies the cell membrane, a selectively permeable barrier that only lets certain substances in and out of the cell.

2. Nucleus

The nucleus behaves like the 'control centre' of the cell, containing the hereditary information (the chromosomes and their constituent DNA). DNA is also found in the chloroplasts and mitochondria.

3. Chloroplasts

The chloroplasts, of which there may be one or several, are the sites of photosynthesis, whereby light energy is captured and converted into a form that can be used by the plant to build simple sugars. They are exclusive to plants.

4. Mitochondria

The mitochondria, like chloroplasts, are concerned with converting energy into a form that can be used to build simple sugars. Rather than using light, however, they use the energy created through the oxidation of sugars, fats and proteins. Mitochondria, therefore, are the main source of energy in the dark. They are much smaller than chloroplasts, and plentiful.

5. Vacuole

The vacuole is a large space within the centre of the cell. It starts off small in young cells and enlarges with age, often causing the other cell contents to be pressed against the cell wall. Some cells have several vacuoles. Their main function is to segregate waste products from the rest of the cell, as they gradually accumulate and sometimes crystallise.

6. Endoplasmic reticulum

The endoplasmic reticulum takes the appearance of many flattened structures layered together. Its surfaces are studded with ribosomes, which are where proteins are made.

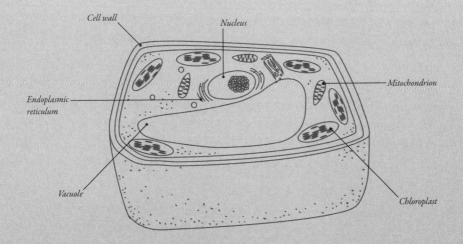

concrete. It gives wood its strength and certainly will not allow wilting if the plant loses water. Cellulose and lignin are believed to be the two most abundant organic compounds on Earth.

The formation of lignin is considered by evolutionary biologists to have been crucial in the adaptation of plants to a terrestrial environment. The theory is that only with lignin could cells be built that are rigid enough to conduct water to any useful height, and against the force of gravity, without the cells collapsing.

Cell division

There is a limit to the size of any cell, partly due to the distance over which the nucleus can exert its controlling influence. For a plant to grow to a much greater size, therefore, cells need to multiply themselves, and this is achieved through division.

Cell division brings about many advantages. It allows cells to specialise, it increases the ability of an organism to store nutrients, it enables damaged cells to be replaced, and it may confer a competitive advantage: large plants have better access to light. There are two types of cell division that living organisms undergo: mitosis and meiosis.

Mitosis
Mitosis is the division of a single cell into two identical new cells; it is responsible for vegetative growth.

Meiosis
Meiosis is a special type of cell division essential for sexual reproduction; it results in four half-cells (gamete cells) rather than two complete cells (as in mitosis). As these cells only contain one set of chromosomes, they are known as 'haploid'. (Most cells contain two sets of chromosomes they are known as 'diploid'.) In higher plants, the gamete cells become either the pollen grains (male cells) or the embryo (female cells). For further information on sexual reproduction, see pp. 110–115.

Mitosis cell division creates daughter nuclei that are genetically identical to the parent cell. Meiosis results in four daughter half-cells.

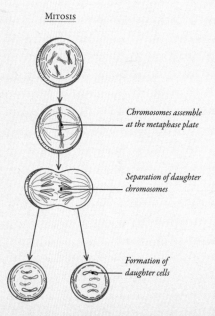

MITOSIS

Chromosomes assemble at the metaphase plate

Separation of daughter chromosomes

Formation of daughter cells

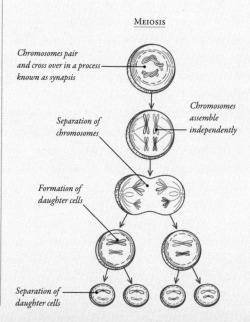

MEIOSIS

Chromosomes pair and cross over in a process known as synapsis

Separation of chromosomes

Chromosomes assemble independently

Formation of daughter cells

Separation of daughter cells

Photosynthesis

The key feature that sets plants apart from almost all other forms of life is their ability to synthesise the building blocks of life using sunlight as an energy source. It is a remarkable biochemical reaction and of such fundamental importance that, were photosynthesis to stop happening, virtually all life on Earth would die.

The overall equation for photosynthesis can be written as follows:

$$6\,CO_2 + 6\,H_2O \rightarrow C_6H_{12}O_6 + 6\,O_2$$

Carbon dioxide = CO_2
Water = H_2O
Glucose = $C_6H_{12}O_6$
Oxygen = O_2

Thus six molecules of carbon dioxide (found in the air) are combined with six molecules of water to produce one molecule of a simple sugar (glucose) plus six molecules of oxygen. Oxygen is therefore a waste product of this reaction; this is another way in which animals rely on plants, since animals need oxygen to breathe, and oxygen is supplied by plants.

How it works

The reaction above does not occur by itself. It requires energy. In nature this energy comes from sunlight, but plants can be made to photosynthesise under the right kind of artificial light. Some horticulturists and farmers exploit this by illuminating plants grown in glasshouses to stimulate growth during periods of low light.

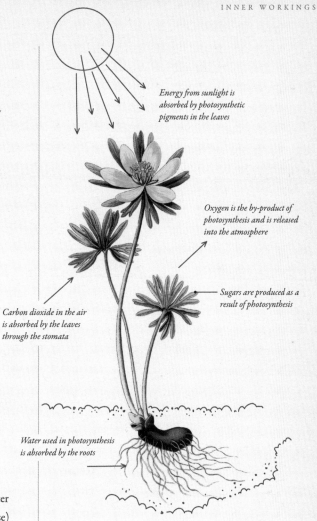

Energy from sunlight is absorbed by photosynthetic pigments in the leaves

Oxygen is the by-product of photosynthesis and is released into the atmosphere

Sugars are produced as a result of photosynthesis

Carbon dioxide in the air is absorbed by the leaves through the stomata

Water used in photosynthesis is absorbed by the roots

It is the chloroplasts that are able to receive and convert light energy. They are filled with a jelly-like substance called stroma. Within the stroma are a number of fluid-filled sacs that contain light-absorbing pigments – the site of the light reaction.

Of all the pigments, green chlorophyll is the most abundant, although other pigments such as carotene (which gives carrots their orange colour) and xanthophyll (yellow) may also be present.

Light reaction

The light reaction begins when light hits pigment molecules, energising the electrons within it. On return to their 'ground' state, electrons can release energy in one of four ways: by emitting heat or light (which can cause phosphorescence), by exciting another electron in another pigment molecule, or by driving a chemical reaction.

In photosynthesis, energised electrons are used to drive two important reactions: the first is the conversion of an ADP molecule (adenosine diphosphate) to an ATP molecule (adenosine triphosphate), and the second is the splitting of water (H_2O) into hydrogen (H) and oxygen (O). This latter process, of which oxygen is the waste product, is a remarkable feat, since water is a very stable molecule. The two free hydrogen atoms combine with a substance called NADP, converting it to NADPH2.

Both reactions cause conversion of light into chemical energy, in the form of ATP and NADPH2. This chemical energy then drives the dark reaction, which takes place in the stroma of the chloroplast.

Dark reaction

The dark reaction is also known as the Calvin cycle, named after one of the people who worked to discover it, Melvin Calvin. Calvin fed plants radioactive carbon and tracked it through the plant in order to make his discoveries. It is called the dark reaction because this stage of photosynthesis is not directly driven by light.

In the Calvin cycle, the first sugars are formed from carbon dioxide by a reaction driven by ATP and NADPH2 molecules. These are converted back to ADP and NADP, and they are recycled in the light reaction. The sugars are the first building blocks of plant life, and they are soon converted into far more complex molecules such as starch – the main storage food of plants. Further reactions with nutrients such as nitrogen lead to the formation of proteins and oils.

Mitochondria and respiration

As an aside, it is worth mentioning the metabolic activities of mitochondria. Like chloroplasts, they are little powerhouses that function to drive living processes. The essential difference is that mitochondria release energy by burning stored fuel rather than converting it from light. The reaction, called respiration, can be simplified as follows:

$$C_6H_{12}O_6 + 6\,O_2 \rightarrow 6\,CO_2 + 6\,H_2O$$

Written down, it is the opposite reaction to photosynthesis – releasing water and carbon dioxide as by-products, rather than using them as metabolites. From it you can see how plants and animals depend on each other: animals breathe in oxygen and release carbon dioxide; plants take in this carbon dioxide to make their own food, with oxygen as a by-product in an endless cycle.

All green life recycles the air that we breathe and makes the planet habitable, which is what makes the Earth's forests so important.

Respiration is the only source of energy in animals, as they do not possess chloroplasts. Plants contain both mitochondria and chloroplasts, so they can respire and photosynthesise. At night, or during low light levels when photosynthesis stops, plants must rely on respiration. The oxygen required for respiration enters the plant through the stomata, so it is not just carbon dioxide that passes through these tiny pores.

Glechoma hederacea,
ground ivy

Plant nutrition

Any gardener will be aware of the need to feed their plants occasionally, particularly those grown in containers, which are entirely dependent on the gardener to keep them supplied with what they need. On the shelves of garden centres are masses of products devoted to the nutrition of plants.

Numerous chemical elements are essential for plant growth, and these are divided into macronutrients and micronutrients. Some of these the plant is able to source from the air (carbon and oxygen), while the great majority are sourced from the soil. Deficiencies in any nutrient often show specific symptoms, and if noticed by the attentive gardener can often be remedied by treating with a specific fertiliser.

Camellias need a special ericaceous feed to provide the specific nutrients they require for healthy growth.

Like most ornamental trees and shrubs, Japanese maples (*Acer* spp.) are best fed with a general, balanced fertiliser, as required.

Macronutrients

Macronutrients are those that are needed in the greatest quantity. They are carbon, oxygen, nitrogen, phosphorus, potassium, calcium, sulphur, magnesium and silicon. Plants rarely have trouble sourcing enough carbon and oxygen, since this is freely available from the air.

Off-the-shelf fertilisers usually list their nutrient content as an N:P:K ratio. N is an abbreviation for nitrogen, P for phosphorus and K for potassium; if, for example, the N value is higher than the P or K value, then it is a fertiliser rich in nitrogen. Some fertilisers also contain extra nutrients, usually special formulations such as those for roses or ericaceous plants (typically containing extra magnesium or iron). Gardeners should be wary of products that do not quote an N:P:K ratio, as their nutrient content cannot be guaranteed. An example might be some dried seaweed products.

Fertilisers containing nitrogen will help produce plenty of green, leafy growth on leaf vegetables, such as spinach.

Nitrogen

Nitrogen (N) is a major plant nutrient. It is vital to plant growth as it is an essential component of all proteins and chlorophyll. If it is deficient, plants fail to grow strongly. In the soil, it is derived from organic matter, and it becomes slowly available to plant roots as this organic matter is decayed by soil micro-organisms into nitrates and ammonium salts. Nitrogen-fixing bacteria living freely in the soil or in root nodules (see chapter 2) are able to shortcut this process by obtaining nitrogen directly from the air.

Nitrates and ammonium salts are highly soluble, so available nitrogen is easily washed out of the soil by overwatering or high rainfall. Flooding, drought and low temperatures can also affect the availability of nitrogen. Deficiency results in stunted or slow growth and pale, yellowish foliage (chlorosis). Sources of nitrogen include well-rotted manure, blood, hoof and horn fertilisers, and ammonium nitrate fertilisers.

Phosphorus

Phosphorus (P) is another major nutrient needed for the conversion of light energy to ATP during photosynthesis and is used by numerous enzymes. It is important for cell division and it is commonly associated with healthy root growth. Legumes have a high demand for phosphorus, but requirements vary between plants. Phosphorus deficiency is uncommon on well-cultivated soils as it is fairly immobile in the soil, but slow growth and dull, yellowish foliage are symptoms. Sources of phosphorus include rock phosphate, triple super-phosphate, bonemeal, and fish, blood and bone.

Potassium

Potassium (K) is the third major plant nutrient needed for photosynthesis, controlling water uptake by the roots and reducing water loss from the leaves. As it also promotes flowering and fruiting it is needed by plants that we specifically grow for their flowers and fruit, and it increases general hardiness. Deficiency causes yellow

High-potash feeds will improve the crop of fruiting plants, such as gooseberries.

or purple leaf tints, and poor flowering and fruiting are symptoms. Tomato fertilisers are probably the most commonly available potassium-rich feeds; sulphate of potash is another.

Sulphur

Sulphur (S) is a structural component of many cell proteins and is essential in the manufacture of chloroplasts, making it vital for photosynthesis. Its deficiency is rarely seen, especially in industrialised countries where sulphur dioxide is often rained down onto the soil from the atmosphere. Flowers of sulphur can be used to lower the pH of soils (see p. 144).

Calcium

Calcium (Ca) regulates the transport of nutrients between cells and is involved in the activation of certain plant enzymes. Calcium deficiency is rare, resulting in stunted growth and blossom end rot – the softening and blackening of the bottom of the fruit. Its concentration in the soil determines acidity and alkalinity (see chapter 6). It can be added to the soil as chalk, limestone or gypsum.

Magnesium

Magnesium (Mg) is essential for photosynthesis, and in the transport of phosphates. Deficiency results in interveinal chlorosis (yellowing between the leaf veins) and it is aggravated by soil compaction and waterlogging. Deficiency is often seen on sandy acidic soils. Foliar sprays of magnesium sulphate or Epsom salts are a useful remedy.

Silicon

Silicon (Si) strengthens cell walls, so improves overall physical strength, health and productivity. Other benefits include improved drought, frost and pest and disease resistance. Many grasses have a high silicon content, which is thought to be an adaptation to deter grazers. Take, for example, the extremely sharp leaf edges of the pampas grass (*Cortaderia selloana*); anybody who has cut their hands on these leaves will not be surprised to learn that silicon is also used to make glass.

Micronutrients

Micronutrients, sometimes referred to as trace minerals, are required in much smaller quantities. Despite this, they are essential for good growth as they play vital parts in many biochemical activities. They include boron, chlorine, cobalt, copper, iron, manganese, molybdenum, nickel, sodium and zinc.

> **BOTANY IN ACTION**
>
> Typical micronutrient deficiencies include lack of iron, which results in interveinal chlorosis; a shortage of manganese, which may result in coloration abnormalities, such as spots on the foliage; and lack of molybdenum, which can cause twisted growth – often seen in brassicas.

Potassium improves flowering and consequent fruiting in tomatoes.

Charles Sprague Sargent
1841–1927

Charles Sprague Sargent was an American botanist with a particular passion for dendrology – the study of trees. He did not receive any formal botany education or training, but possessed excellent botanical instincts.

Sargent's father was a wealthy Boston banker and merchant, and Sargent grew up on his father's Holm Lea estate in Brookline, Massachusetts. He went to Harvard College and after graduation enlisted in the Union Army, and saw service during the American Civil War. After the war, he travelled throughout Europe for three years.

On his return to America, Sargent started his long horticultural career by taking over the management of the family's Brookline estate, much inspired by Horatio Hollis Hunnewell. Hunnewell was an amateur botanist and one of the most prominent horticulturists in America in the nineteenth century. With the help of Hunnewell and his unique instruction, Holm Lea estate was converted into a living landscape without geometric designs and flowerbeds, but with a more natural look, including mass plantings of trees and shrubs. It would soon develop a world-class collection of rhododendrons and majestic trees.

Sargent concentrated much of his time on developing his arboretum. He worked with Frederick Law Olmsted, popularly considered to be the father of American landscape architecture, and was heavily

Charles Sargent published several works of botany and the abbreviation Sarg. is attributed to plants he described.

involved with every aspect – from the overall master planning to much smaller details, including selecting which individual trees to plant.

Sargent soon became regarded as a leading dendrologist. He started to write about trees and shrubs and was widely published. He also became the linchpin in America's rhododendron story. His skills and knowledge were sought nationally, especially with regards to the conservation of American forests, particularly the Adirondack and Catskill forests in New York State. He was even elected chairman of a commission to help preserve the Adirondacks.

In 1872, Harvard University decided to create an arboretum after James Arnold left more than $100,000 to Harvard for 'the promotion of agricultural or horticultural improvements'. Professor Francis Parkman, then Professor of Horticulture at Harvard's school of agriculture and horticulture, the Bussey Institution, suggested that Sargent should be heavily involved in its creation.

Sargent, along with Olmsted, undertook a massive job planning and designing the arboretum, as well as working on the funding to ensure its continuing success. By the end of that year, Sargent was appointed the first Director of what was named Harvard's Arnold Arboretum, a position he held for 54 years until he died. During that time, it grew from the original 50 ha (120 acres) to 100 ha (250 acres). Sargent also continued his own research and writing.

Besides collecting plants and specimens, Sargent also amassed a large collection of books and journals for the Arnold Arboretum Library. The collection grew from nothing to more than 40,000 books during his time. Most of these were purchased at Sargent's expense and by the time of his death he had donated his entire library to the Arboretum as well as providing finance for the upkeep of the collection and the purchase of further materials.

Later he became Professor of Arboriculture at Harvard. He was also appointed Director of the Botanic Garden in Cambridge, Massachusetts, a garden that has long since disappeared.

He regularly wrote about his passion for trees and published several works of botany. In 1888 he became editor and general manager of *Garden and Forest*, a weekly journal of horticulture and forestry. His publications include: *Catalogue of the Forest Trees of North America*; *Reports on the Forests of North America*; *The Woods of the United States, with an Account of their Structure, Qualities, and Uses*; and the 12-volume *The Silva of North America*.

After his death, Massachusetts Governor Fuller commented: 'Professor Sargent knew more about trees than any other living person. It would be hard to find anyone who did more to protect trees from the vandalism of those who do not appreciate the contribution that they make to the beauty and wealth of our nation.'

Rhododendron ciliatum,
rhododendron

Charles Sargent amassed a world-class collection of rhododendrons at his Holm Lea estate and was important in establishing non-native rhododendrons in America.

Sadly, after Sargent's death, his large plant collection had to be sold and was broken up and bought by individual plant collectors and breeders. Plants named after him include *Cupressus sargentii*, *Hydrangea aspera* subsp. *sargentiana*, *Magnolia sargentiana*, *Sorbus sargentiana*, *Spiraea sargentiana* and *Viburnum sargentii*. The standard author abbreviation Sarg. is used to indicate him as the author when citing botanical names of plants.

Picea sitchensis,
sitka spruce

Charles Sargent was famous for his work with trees and was instrumental in setting up the Arnold Arboretum.

Nutrient and water distribution

Primitive plants, such as algae, are able to rely on diffusion of substances within their cells as a means of nutrient distribution, from areas of high concentration to areas of low concentration. As plants gain in complexity, however, diffusion alone is not sufficient, so specialised transport (or vascular) systems are required to transport water and nutrients from one part of a plant to another.

In chapter 2 (see pp. 64–65), the two types of transport vessels were introduced: xylem and phloem. The xylem transports water and soluble mineral nutrients from the roots throughout the plant, and the phloem mainly transports organic substances created by photosynthesis and other biochemical processes.

Their combined action brings water and nutrients to all the living tissues of the plant, while exchange through the stomatal pores usually relies on diffusion, with gases moving from areas of high to low concentration.

Transport through the xylem

If a capillary tube is dipped into water, water will spontaneously rise up into it due to its high surface tension. Xylem vessels work along these lines, but even in the finest xylem vessels (which are microscopically thin) water would only rise to about 3 m (10 ft) by this method of capillary action. Other forces must therefore be acting on the xylem to enable water to be pulled through a plant the size of a tree.

The cohesion–tension theory states that it is the evaporation of water from the leaves (transpiration) that is responsible for pulling water from the

Sequoia sempervirens,
coast redwood

Even large trees, such as the giant redwood, easily move water and nutrients around themselves through the vascular tissue.

roots. As water leaves the vessels in the leaf, a tension is set on the water that is transferred all the way down the stem to the roots, much like water being sucked through a straw. Inward pressures on the xylem can be incredibly high, and to prevent the xylem collapsing it has special spiral- or ring-shaped reinforcements in the cell walls.

Transpirational pull can generate enough force to lift water hundreds of metres to a tree's highest branches, and this water can move incredibly quickly, at speeds of up to 8 m (26 ft) per hour.

Critics of this theory point to the fact that any break in the column of water should stop its flow. This is not observed, however, and the reason is thought to be because water can flow from one vessel to another, bypassing any air-locks along its way.

Transport through the phloem

In contrast to xylem vessels, phloem is composed of living tissue, and it transports a solution rich in sugars, amino acids and hormones throughout the plant. This process is called translocation, and it is not properly understood.

Phloem tissue consists of sieve tube cells and companion cells. Any theory that attempts to explain how translocation works must take this anatomy into account, and it must also be able to answer important questions. How does the phloem manage to transport large quantities of sugars? Why are only small amounts of phloem present in a plant? How do substances move both up and down the phloem?

The needle-like mouthparts of aphids are able to puncture the soft new growth of plants, and they tap directly into the phloem. Scientists exploit this mechanism by separating an aphid from its mouthparts while it is gorging on a plant's sap. The sap that exudes from the separated mouthpart can be collected and analysed. This is one way that scientists are able to study the movement of substances around the phloem.

The mass flow hypothesis was first proposed in 1930, and it attempts to explain translocation by 'sources' and 'sinks'. Phloem sap moves from sugar sources (areas of high concentration) to sugar sinks (areas of low concentration). Areas of sugar concentration constantly vary in a plant, with the leaves becoming sources during periods of active photosynthesis, and tubers becoming sources during periods of low growth or dormancy.

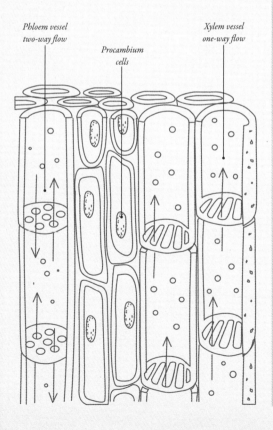

A longitudinal cross section of a typical vascular bundle, showing the general layout of the xylem and the phloem, and the direction of flow of substances within them.

Plant hormones

In the absence of a nervous system, the control of growth in plants must be entirely regulated by chemical signals. Five main groups of these chemical signals are recognised, although more are likely to be discovered, and they are known as plant hormones.

A plant hormone is an organic compound synthesised in one part of a plant and translocated to another part, where in very low concentrations it causes a physiological response. Such responses may be promotive or inhibitive (ie, they may cause a plant to do more of something, or less).

Avena sativa, cultivated oat

The idea that plant development is influenced by special chemicals is not new and was proposed over one hundred years ago by Julius von Sachs, a German botanist. Since the concentrations of hormones are so low, however, it was not until the 1930s that the first plant hormone was identified and purified.

Auxin

In 1926, Frits Went found evidence of an unidentified compound that caused curvature of oat shoots towards the light. This phenomenon is known as phototropism and gardeners have the hormone auxin to thank when their plants lean towards the light! Even though its action is still poorly understood, other auxin hormones are known to influence bud and leaf formation, and leaf drop. Hormone rooting powders contain auxins because of their influence on root production. This is useful for gardeners who take cuttings.

Gibberellin

In the 1930s, Japanese scientists isolated a chemical from diseased rice plants that grew excessively tall and were unable to support themselves. The disease is caused by a fungus called *Gibberella fujikuroi*, which was found to overload the rice plants with a chemical that was named gibberellin. To date many gibberellins have since been discovered and they are found to be important in promoting cell elongation, seed germination and flowering.

Cytokinins

In 1913, Austrian scientists discovered an unknown compound present in vascular tissues that stimulated cell division and subsequent cork formation and wound healing in cut potato tubers. This was the first evidence that plants contained compounds that could stimulate cell division (cytokinesis). This group of plant hormones are now called cytokinins, and they serve many functions in plant growth.

Abscisic acid

The role of abscisic acid in the fall of leaves (leaf abscission) led to the naming of this plant hormone. It often gives plant organs a signal that they are undergoing physiological stress. Among those stresses are lack of water (the chemical signal is sent from the roots to the stomata, causing them to close), salty soils and cold temperatures. The production of abscisic acid, therefore, causes responses that help protect plants from these stresses. Without it, for example, buds and seeds would start to grow at inappropriate times.

Pisum sativum,
pea

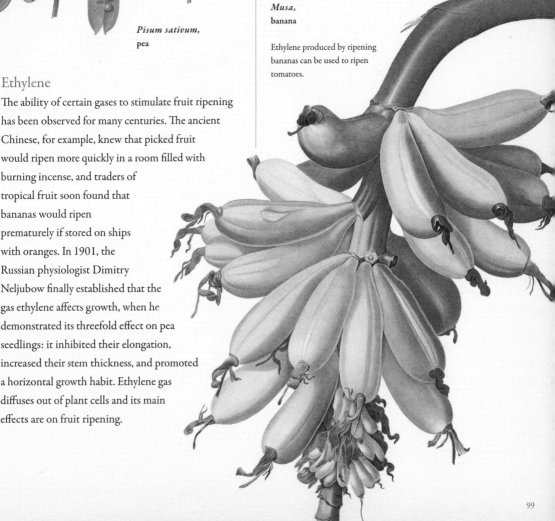

Musa,
banana

Ethylene produced by ripening bananas can be used to ripen tomatoes.

Ethylene

The ability of certain gases to stimulate fruit ripening has been observed for many centuries. The ancient Chinese, for example, knew that picked fruit would ripen more quickly in a room filled with burning incense, and traders of tropical fruit soon found that bananas would ripen prematurely if stored on ships with oranges. In 1901, the Russian physiologist Dimitry Neljubow finally established that the gas ethylene affects growth, when he demonstrated its threefold effect on pea seedlings: it inhibited their elongation, increased their stem thickness, and promoted a horizontal growth habit. Ethylene gas diffuses out of plant cells and its main effects are on fruit ripening.

Iris ensata, Japanese iris

Chapter 4

Reproduction

Nobody really knows who first discovered sex in plants, but it is generally credited to the German botanist Rudolf Jakob Camerarius in his book *De sexu plantarum epistola*, published in 1694. At this time, scientists were gradually becoming open to the idea that flowers had male and female parts and that sexual reproduction was occurring between them.

The success of plants, however, is all down to their magnificent ability to proliferate themselves. This can be by sexual reproduction or by vegetative reproduction. The latter is where bits of plants regrow from other bits of plants, and gardeners are constantly exploiting this ability to make new plants with the minimum of difficulty.

To a gardener, a farmer, a plant breeder or an agricultural scientist, the readiness of plants to reproduce is grist to their mill. From seed to shoot, to cutting to offset, we may think of ourselves as their masters; in fact, we are nothing of the sort. We just play our own part in their life cycles, keeping them going, using them as we wish, but perpetuating them from generation to generation. Any gardener faced with a border full of weeds must accept the fact that plant life goes on . . . with or without us.

Vegetative reproduction

This method of asexual reproduction (reproduction without sex) is common to all plants. It is a way that plants can create new individuals from vegetative growth material, such as stems, roots and even leaves. The resulting new plants are genetically identical to the plant they arise from; they are clones.

Specialised structures often develop for the purpose of vegetative reproduction, such as runners and rhizomes (see chapter 2). Storage organs like rhizomes, bulbs and corms are also capable of vegetative reproduction, increasing themselves annually underground. Gardeners often use these as a source of new plants, and potato tubers are a good example. A healthy potato plant may produce about half a dozen sizeable tubers, which can all be replanted to create new plants. All the autumn bulbs seen for sale in garden centres will be the result of vegetative reproduction on a huge scale.

It is the presence of meristematic tissue in most parts of a plant that gives them the potential of generating a whole new plant. This ability is called totipotency – the ability of cells to produce new plants. In theory, any part of a plant could probably be encouraged to regrow so long as it contains meristems, but experienced gardeners know that for different plants, they are likely to have more success with cuttings from certain parts of a plant. For example, with Japanese anemones (*Anemone* × *hybrida*), root cuttings are the preferred method, while lavenders (*Lavandula*) are more likely to strike from stem cuttings.

In recent years, scientists have developed micropropagation techniques, by which plants can be encouraged to grow from cell cultures of meristematic tissue in the laboratory. While micropropagation is outside the realm of most gardeners, it is a technique that has revolutionised the commercial production of certain plants, such as hostas. It is much faster than dividing the plants, particularly those that are slow to grow on, and it allows huge numbers of new plants to be grown from one sample. It has made many types of plants both affordable and plentiful.

Vegetative reproduction and the gardener

In cultivation, vegetative reproduction is sometimes preferred as it can be exploited to maintain desirable characteristics that would probably be lost or diluted through sexual reproduction. For the mass production of garden plants, it is widely used.

Vegetative reproduction is often preferred by gardeners simply because it is easier than growing from seed, particularly if a plant is slow or reluctant to produce seed. Some cultivated plants are actually

Lavandula stoechas,
French lavender

Reproduction

Rosa 'Duc d'Enghien', Bourbon rose

Plant offsets

Offsets are young plants or plantlets that develop while still attached to their parent, either above or below ground. They are easily detached and grown on. They are seen on plants such as saxifrages, sempervivums and phormiums. Initially, offsets have few roots of their own and are dependent on the parent plant for food; roots usually develop towards the end of the first growing season.

Some monocotyledonous plants, such as *Cordyline* and *Yucca*, produce shoots and young plants from their roots, which are also referred to as offsets. These can be removed by carefully scraping away the soil at the base of the plant and removing the offset with a sharp knife, preferably with a section of root attached.

Cordylines (such as *C. stricta*) can be propagated by carefully removing offsets (young plants), produced from their roots at the base of their stems.

incapable of producing seeds, such as roses with double flowers, which have been bred so that all the sexual parts have been converted into petals. In such cases, vegetative propagation is the only method by which plants can be increased.

Division

The commonest method of vegetative reproduction for herbaceous perennials is division, whereby the crown of the plant is pulled or forced apart to produce two or more plants. For the gardener, nothing could be simpler as no special knowledge or materials are needed beyond a spade and fork, although the tough crowns of some grasses and bamboos may need to be chopped or sawn.

Division can also rejuvenate tired and worn-out plants that are not performing well. In such instances, the older growth in the centre of the crown is usually discarded. The best time to divide most herbaceous perennials is immediately after flowering or, for late-flowering plants, in autumn or the following spring.

Runners

Runners are a type of offset, generally comprising horizontal stems that grow from the main plant and then creep along the ground. They produce plantlets along their length or at their ends, and strawberries (*Fragaria* × *ananassa*) are a typical example.

If you propagate by this method you should thin out the runners to encourage those that remain to become stronger. It is better to have a few larger plants than lots of small ones. Pin down the plantlet with a wire hoop in well-prepared soil or into a small pot of compost. Sever the connecting stem when the plant has fully rooted.

Some plants use runners and stolons, as well as rhizomes and rooting stems (see chapter 2) to colonise huge areas. The very invasive and difficult-to-eradicate rhizomes of Japanese knotweed (*Fallopia japonica*) make this plant one of the world's most 'successful' weeds, the persistent rhizomes of horsetail (*Equisetum*)

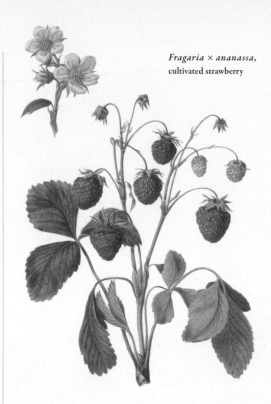

Fragaria × *ananassa*,
cultivated strawberry

Fallopia japonica,
Japanese knotweed

are a problem in many gardens, and the rooting, arching stems of the common bramble (*Rubus fruticosus*) soon colonise new ground. While the continually spreading habit of many plants and garden weeds might be a bane to the gardener, in many ecosystems the root structures of these plants are very important in preventing soil erosion. Marram grass (*Ammophila*), for example, is very important in stabilising sand dunes and preventing the deterioration of coastlines.

Grafting

One part of a plant can be transplanted onto another in a process known as grafting. The two parts will eventually function as one plant. The top half of the graft is called the scion, and the lower half is called the rootstock. Grafting is often used to reproduce plants that are difficult to propagate by other methods, but it is also commonly used by nurserymen as a fast way to produce more stock.

The technique also has the advantage of combining the desirable features of both plants. A rootstock can be chosen to improve a plant's tolerance of soil type or pest resistance, and the scion chosen for its ornamental or fruiting qualities. Grafting is often performed on fruit trees, with rootstocks chosen for vigour (for example: dwarfing or semi-dwarfing) and scions chosen for the cultivar. Thus, you may have apple 'Lord Lambourne' grown on a very dwarf 'M27' rootstock if it is required as a small, step-over tree.

Grafting is also performed on some fruiting vegetables, such as tomato and aubergine plants. While this is not a new technique, it is becoming quite common now to find grafted plants for sale by mail-order nurseries. The advantage of grafted plants is that they have rootstocks chosen for their vigour and high resistance to soil-borne pests and diseases. The scions are chosen for their delicious fruit, and when combined with a vigorous rootstock, yields can be much higher.

Cuttings

The taking of cuttings by gardeners is a form of vegetative reproduction. Cuttings are part of a plant that can be encouraged to grow. The fact that plants are able to regenerate this way suggests that it is a trait evolved in response to environmental opportunities.

Many trees and shrubs that grow beside water, such as willows (*Salix*), are able to regenerate from hardwood cuttings, which are cuttings taken in winter when the wood is mature and leafless. Hardwood cuttings are easy to strike, as all that is needed is to push them into the soil and let nature do the rest. In the wild, winter storms and flooding may tear branches from waterside trees, causing them to be washed away. If a plant can regenerate from these 'cuttings' when they are finally deposited on some distant bank, then this will enable a plant to colonise a new area.

Growing new plants from short sections of stem or root is a very useful and versatile method of vegetative propagation. Many different plants can be propagated this way, including trees, shrubs, climbers, roses, conifers, herbaceous perennials, fruit, herbs, indoor plants and half-hardy perennial bedding plants. The aim of propagation from stem cuttings is to initiate and develop a good root system on the stem from adventitious cells.

Drying out and infection are two problems that cuttings face before they take root. Gardeners can help the cutting by reducing the total leaf area (sometimes a few of the leaves on the cuttings are removed), by keeping the cuttings well misted and by placing them in a partially shaded area.

Learning to take cuttings takes time and experience, gaining knowledge from others and your own successes and failures. Every different species has its own particular requirements, and personal judgement plays a huge role. Gardeners usually take many cuttings at any one time, knowing that some of them may die before they take.

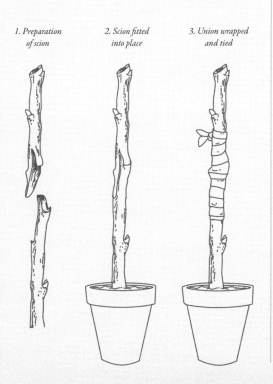

1. Preparation of scion

2. Scion fitted into place

3. Union wrapped and tied

Hydrangea macrophylla,
hydrangea, hortensia

Improving the rooting success of cuttings

Some plants are more difficult to root than others, but it is possible to speed up and assist rooting through different approaches. Hormone rooting powders or gels may help, but they are not miracle cures and may have no beneficial effect; overdosed cuttings may even die.

Some hard-to-root plants respond well to wounding the stem by removing a thin, vertical sliver of bark up to 2.5 cm (1 in) long at the base of the cutting and then dipping the wound in hormone rooting compound.

Cuttings of large-leaved plants, such as *Hydrangea*, benefit from having their leaves cut in half horizontally, which reduces the leaf area and so reduces water loss and wilting. Other plants root better when taken with a heel, which is a tail of bark from the main stem.

The most common types of cuttings

There are four main types of stem cuttings – softwood, greenwood, semi-ripe and hardwood. These are taken at different times of the year, depending on when the required plant material becomes available and, consequently, from different types of stem growth.

The vast majority of stem cuttings are nodal, i.e. the base of the cutting is cut immediately below the swelling of the node. The nodal area contains a high proportion of adventitious cells that have the ability to form roots and a concentration of hormones to stimulate root production. The tissue just below the node is also usually harder and more resistant to fungal diseases and rots.

Softwood cuttings

Softwood cuttings are taken from the most immature part of the stem, produced continuously during the growing season at the growing tips. Although they can be taken at any time during the growing season, they are more commonly taken in spring, which gives them time to establish before winter.

Because of their soft nature, softwood cuttings are often the most difficult to keep alive. Fortunately, this growth has the greatest potential of all stem types to produce roots, because of its youth and vigour.

The rapid growth rate does have one drawback: cuttings lose a lot of water. Once they have started to dry out and wilt they will never root, so it is vital that the gardener takes precautions. Only collect as much material as can be dealt with immediately, and collect it in a damp plastic bag, sealed to maintain humidity. A piece of damp cotton wool placed in the bag also helps.

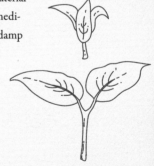

Softwood cutting

Greenwood cuttings

Greenwood cuttings are similar to softwood cuttings, but are taken from the tips of leafy stems later in the year, usually from late spring to midsummer.

Semi-ripe cuttings

Semi-ripe cuttings are taken from midsummer until mid-autumn, when stems start to ripen and harden. The base of the cutting should be hard, while the tip is still soft.

Greenwood cutting

As they are thicker and harder than softwood cuttings and have a greater store of food reserves, they are much easier to keep alive. But because they are usually quite leafy, they have the same issue with water loss and wilting.

Semi-ripe cutting

Hardwood cuttings

Hardwood cuttings are the easiest as they have no leaves to rot and plenty of stored food reserves. This is the perfect method for a number of deciduous trees, shrubs, roses and soft fruit. Cuttings are taken from fully ripened wood, when the plant is fully dormant after leaf fall. As hardwood is the oldest wood and has the least vigour, always select the stems that show the strongest growth.

Although this type of cutting may be slow to develop roots and shoots, it is usually successful. The cuttings will have rooted within 12 months, when they can be lifted and potted up or planted out.

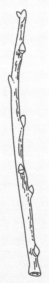

Hardwood cutting

Root cuttings

Root cuttings are restricted to those plants that have the ability to produce adventitious shoots from the roots. The list of plants that can do this is much smaller than for stem cuttings, but it does include a number of popular garden plants such as Japanese anemones (*Anemone* × *hybrida*), stag's horn sumach (*Rhus typhina*), *Geranium sanguineum* and *Primula denticulata*. Their long, fleshy roots are severed during late winter and planted vertically in the ground, ideally in a cold frame.

Leaf cuttings

Leaf cuttings can be made from a number of houseplants such as African violets (*Saintpaulia*), mother-in-law's tongue (*Sansevieria trifasciata*) and many begonias. Less well known are the outdoor plants that can be propagated by such means, such as snowdrops (*Galanthus*) and snowflakes (*Leucojum*). Leaves should be cut or trimmed, partially buried in cutting compost, and then left to root in a closed case shaded from the sun.

Galanthus elwesii, **greater snowdrop**

Luther Burbank
1849–1926

Luther Burbank was one of the leading American horticulturists and a pioneering agricultural and horticultural plant breeder. He devoted his entire time to breeding and creating new plants and succeeded in producing more useful new cultivars of fruit, flowers and vegetables than any other person. One of his main goals was to increase the world's food supply by manipulating plant characteristics.

Burbank was born in Lancaster, Massachusetts, and grew up on the family farm where he particularly enjoyed growing plants in his mother's garden. He was more interested in nature and how things grew than in playing. When his father died, when Burbank was 21, he used his inheritance to buy 6.9 ha (17 acres) of land and started work on potato breeding. It was here he developed the Burbank potato and by selling the rights to it, earned $150, which he used to travel to Santa Rosa, California.

In Santa Rosa he bought 1.6 ha (4 acres) of land that became his outdoor laboratory, where he conducted his famous plant hybridisation and cross-breeding experiments that would, in a short spell of time, bring him world renown. He produced multiple crosses between foreign and native plants to provide seedlings that he would assess for worthiness. Often these were grafted onto fully developed plants so that he could assess them for hybrid characteristics much faster. Within a few years, he was conducting so many experiments that he needed to expand. He bought more land near Santa Rosa, which became known as Luther Burbank's Gold Ridge Experiment Farm.

During this career, Burbank introduced more than 800 new varieties of plants, including more than 200 types of fruit (particularly plums), many vegetables and nuts, and hundreds of ornamental plants.

These new plants included a spineless cactus that provided forage for livestock in desert regions and the plumcot – a cross between a plum and an apricot. Other famous fruit introductions include the Santa Rosa plum (which as late as the 1960s, accounted for more than one-third of California's commercial plum harvest), peach 'Burbank July Elberta', nectarine 'Flaming Gold', freestone peaches, strawberry 'Robusta' and a white blackberry known as the iceberg white blackberry or snowbank berry.

Luther Burbank was a pioneering plant breeder, famous for introducing more than 800 plants, many of which are still grown today.

But it was not only fruit that interested Burbank: he was also responsible for numerous ornamental plant introductions. The most famous of these is probably the Shasta daisy (*Leucanthemum* × *superbum*), which Burbank produced as a hybrid between *Leucanthemum lacustre* and *Leucanthemum maximum*.

Burbank named some of his new cultivars after himself, such as *Canna* 'Burbank', and other plants have been named after him since, including *Chrysanthemum burbankii*, *Myrica* × *burbankii* and *Solanum retroflexum* 'Burbankii'.

Despite his massive breeding programmes and introduction of numerous commercially desirable plants, Burbank was often criticised as being scientific. But he was more interested in results than in pure research.

Burbank wrote or co-wrote several fascinating books on his work, which give an insight into the man and the mammoth task he undertook. They include the eight-volume *How Plants Are Trained to Work for Man*, *Harvest of the Years* and the 12-volume *Luther Burbank: His Methods and Discoveries and Their Practical Application*.

Burbank left behind him a fabulous legacy, not only all the new plants he created. In 1930, the USA's Plant Patent Act – a law that made it possible to patent new cultivars of plants – was passed, helped in a big way by Burbank's work. The Luther Burbank Home and Gardens, in Santa Rosa, is a National Historic Landmark and the Gold Ridge Experiment Farm is listed in the National Register of Historic Places. He was inducted into the National Inventors Hall of Fame in 1986. In California, his birthday is celebrated as Arbor Day and trees are planted in his memory.

Prunus domestica, plum

Illustrations by Alois Lunzer depicting plum cultivars 'Abundance', 'Burbank', 'German Prune' and 'October Purple'.

A cultivar that originated from Burbank's original potato, possessing russet-brown skin, became known as the 'Russet Burbank' potato. This potato is the predominant potato used in food processing and is commonly used by fast food restaurants to produce French fries. It was exported to Ireland to help the Irish people recover from the Irish Potato Famine.

The standard author abbreviation Burbank is used to indicate him as the author when citing botanical names of plants.

> 'I SEE HUMANITY NOW AS ONE VAST PLANT, NEEDING FOR ITS HIGHEST FULFILLMENT ONLY LOVE, THE NATURAL BLESSINGS OF THE GREAT OUTDOORS, AND INTELLIGENT CROSSING AND SELECTION.'
> Luther Burbank

Sexual reproduction

Evolutionary success in a species requires a balance between conservation of favourable genetic characteristics and variation, which is necessary for plants to adapt to their often-changing environment.

Sexual reproduction results in a degree of variation, as the offspring that result will show a mixture of characteristics inherited from both parents. Mutations can also arise during plant growth, and contribute to genetic variation.

As discussed in chapter 2, sexual reproduction in flowering plants can only come about with the transfer of male pollen to the stigma, by a process known as pollination. The growth of a pollen tube through a flower's style, followed by a meeting of sperm and egg cells, results in fertilisation.

Pollination

The transfer of pollen is usually carried out by wind or by animals, and the mechanism used will be the result of millions of years of adaptation of a plant to its environment. Pollination methods vary between plants and are extremely diverse.

It is easy to tell whether a flower is wind- or animal-pollinated, since the flowers show distinctive characteristics. Wind-pollinated flowers, for example, are inconspicuous, with exposed anthers and stigmas. Animal-pollinated flowers are usually showy, with colour, scent and nectar to attract and reward a visitor.

The petals of animal-pollinated flowers sometimes have patterns that guide pollinators to the source of the nectar or pollen. These guides may only be visible under ultraviolet light – visible only to insects. Scent is another powerful attractant, as it can travel a long way. Many winter-flowering plants produce highly scented flowers in order to attract pollinators, which may be sparse at that time of year. Flowers pollinated by night visitors, including bats and moths, are likely to be very smelly but not very showy.

Plants pollinated by wind have much less showy flowers and they are often very small. Grasses are a perfect example, but some trees such as birches (*Betula*) and hazels (*Corylus*) release pollen to the wind from their catkins. They produce vast amounts of small, light and non-sticky pollen that can be carried even in the lightest breeze (often a poor time of year for sufferers of hay fever). The pistils that catch the wind-borne pollen are long, with feathery, very sticky stigmas.

Betula alnoides, alder-leaf birch

Birches produce pollen-bearing catkins. Even light breezes blow the pollen to the female stigmas.

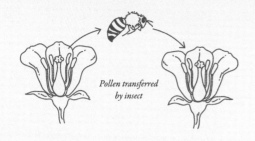

Pollen transferred by insect

Pollen transferred by wind

In sexual reproduction, pollen can either be actively transferred by pollinators or passively transferred by air currents.

It is thought that wind-pollinated trees and shrubs release their pollen when they are not in leaf so that the leaves cannot hinder the transfer of pollen. Even though the individual pollen grains tend to be small and therefore low in nutritional value, some insects will gather it when other pollen is scarce.

Specialised pollination mechanisms

Some highly evolved flowers are very specialised, with hidden anthers and stigmas that can only be reached by pollinators with the appropriate size, shape and behaviour. Bee pollination is probably the best-studied example, as in foxgloves (*Digitalis*) or bluebells (*Hyacinthoides non-scripta*).

Primitive flowers like buttercups (*Ranunculus*) are very open to any visitor, and therefore attract a variety of generalist feeders. These types of flowers have to produce a lot of pollen, much of which is wasted. The most highly specialised flowers attract just a single species of pollinator. They are more efficient as they have to produce less pollen, since the chances of it reaching the stigma of a plant of the same species are much higher. The downside is that these plants are very susceptible to the loss of the pollinator, which may move on due to environmental change.

Mirror orchids (*Ophrys speculum*) are one such example, with flowers that look very much like a female bee, even emitting a similar pheromone (chemical signal) to the one released by the female bee. As a result, male bees copulate with the flowers, thus pollinating them. This mechanism is so specialised in the Cyprus bee orchid (*Ophrys kotschyi*) that only one species of bee pollinates it. In such instances, the plant is not wasting its resources by manufacturing a reward, but it is left very vulnerable to change.

The use of mimicry to attract pollinators is seen in many flowers. The most revolting must be the dead horse arum (*Helicodiceros muscivorus*), which grows

Ranunculus asiaticus,
Persian buttercup

near seagull colonies in Corsica and Sardinia. Their large, greyish-pink spathes resemble rotting meat, and the smell that comes from them is equally revolting. They flower during the gull's breeding season, when the colony is awash with dead chicks, rotting eggs, droppings and uneaten fish scraps. Millions of flies are attracted to the detritus, and the flowers fool many of the flies into pollinating them. In the garden, bad-smelling plants that function in a similar way include lords-and-ladies (*Arum maculatum*), the dragon arum (*Dracunculus vulgaris*) and the skunk cabbage (*Lysichiton americanus*).

Dracunculus vulgaris,
dragon arum

The flowers of dragon arum emit a smell resembling rotting meat that attracts pollinating flies.

Viola riviniana,
common dog violet

Cross-pollination and self-pollination

Readers can be forgiven at this point for asking a simple question: what's to stop a flower from pollinating itself? The anthers and stigmas are very close together in a flower, so this must happen all the time, thereby short-circuiting a plant's attempts at pollinating flowers of other plants (cross-pollination).

The answer is that it does happen. It is called self-pollination, and it may occur between two flowers on the same plant or within a single flower. But rather cleverly, plants have evolved mechanisms that control the extent of cross- and self-pollination. *Viola riviniana*, for example, sometimes produces very small flowers which remain closed, ensuring self-pollination takes place. This helps to make economical use of resources by making certain pollination occurs without having to produce the large flowers and abundant pollen associated with cross-pollination.

Of course, plants do not actively make these decisions themselves. The strategy a plant adopts will be a complex result of interaction with its environment and its own physiology. For example, *Impatiens capensis* resorts to a closed-flower self-pollination strategy in response to heavy grazing.

Cross-pollination is also impossible if pollination occurs before the flower is even fully developed. In the common vetch (*Vicia sativa*) pollination occurs at flower-bud stage. This is a trait common in beans and peas, which is why F1 hybrids of these plants (see pp. 120–121) are rarely seen.

In some flowers, the anthers and stigmas are positioned so that they are not in easy reach of one another. This reduces the chance of self-pollination. Members of the primrose genus (*Primula*) bear two different types of flower: pin and thrum. The pin form has a prominent stigma and anthers recessed into the floral tube. Conversely, the thrum form has exposed anthers and a shorter stigma at the bottom of the tube. This makes it easier for an insect to transfer pollen from one type to the other, and the effect is a much greater rate of cross-pollination. Pin and thrum flowers are also seen in pulmonarias.

Some plants also have male and female forms. For example, most cultivars of the common holly (*Ilex aquifolium*) are either male or female. Such plants are said to be dioecious. It is only in dioecious plants that cross-pollination is guaranteed.

The male and female parts of a flower may also mature at different times, shifting the balance towards cross-pollination. The anthers of the foxglove (*Digitalis purpurea*) mature before the stigma is ready to receive pollen, although in other plants it may be that the stigmas mature before the anthers.

In foxgloves, the flowers are produced along an upright spike, with the lower flowers maturing first. Who hasn't enjoyed a delightful moment watching a bee as it forages from flower to flower, in an upwards direction? Bees forage in this direction, visiting the most mature flowers first (those with receptive stigmas), gradually moving up to those flowers that are producing pollen. The pollen sticks to the bee, and when it reaches the topmost flower it then flies on to the next plant. The first flower that the bee reaches after this is at the bottom of the spike, and thus the pollen is received and cross-pollination achieved.

Common as lawn weeds, the tiny, wind-pollinated flowers of the plantains (*Plantago*) also ripen in succession like those of foxgloves, except that the flowers at the top of the spike mature first. Since the pollen is gathered mainly at the bottom part of each flowerhead, there is a reduced chance of it falling onto the receptive stigmas of the same plant as these are higher up. The pollen grains are instead blown away to cross-pollinate other plants.

Ilex aquifolium,
common holly

Self-incompatibility

Gardeners who grow tree fruits may wonder about the issues of self-incompatibility and pollination groups. Apple trees, in particular, are noted for this, and accepted advice is that if you are going to plant an apple tree, you will need to plant another in the same pollination group if fruit is to be set.

The reason for this lies in incompatibility. Sometimes it is not enough for the pollen of one flower to land on the stigma of another; there may be compatibility issues. In the case of most apples, self-incompatibility is the issue, with successful pollination of an individual tree using its own pollen being impossible.

In such cases, pollination may actually occur, but the incompatibility issue arises once fertilisation of the embryo is attempted. In the case of cocoa plants (*Theobroma cacao*), fertilisation may actually take place but is soon aborted, owing to incompatibility.

For this reason, apple, cherry, pear, plum, damson and gage trees are all divided into pollination groups, depending on flowering time and compatibility. Some particular varieties, such as the 'Victoria' plum, are self-compatible, so only one specimen needs to be grown for successful fruiting. Gardeners must consider pollination groups when planning a fruit garden.

Pollen formation

Each stamen consists of an anther (usually two-lobed), in which pollen sacs produce pollen, and a filament, which contains vascular tissue and connects the anther to the rest of the flower.

Within each pollen sac, pollen mother cells divide by meiosis (see p. 88) to form the haploid pollen-grain cells. Each pollen grain will develop a thick, sculptured wall characteristic of that particular species.

The pollen-grain cell then divides by mitosis to form a pollen tube nucleus and a generative nucleus. The contents of the pollen grain are now said to be of the male gametophyte generation. As mentioned on p. 25, this has become extremely reduced in flowering plants, and here it is – just two tiny half-cells.

Theobroma cacao, cocoa

BOTANY IN ACTION

Under an electron microscope scientists are often able to tell different species apart by the sculpture of their pollen grains. Pollen grains can be preserved for thousands of years in soil, and by studying the pollen content of different layers of soil, scientists are able to make observations about how a region's flora has changed over time.

Development of the ovule and embryo

Each pistil consists of at least one stigma, style and ovary. Within an ovary, one or more ovules develop, each attached to the ovary wall at a place called the placenta by a short stalk called a funicle. Food and water pass through to the ovule via the funicle.

The main body of the ovule is the nucellus, enclosed and protected by the integuments. Within the nucellus, an embryo sac mother cell develops to form the mature embryo sac. The embryo sac mother cell undergoes meiosis and just one of these haploid cells grows on to form the mature embryo sac.

Growth of the haploid embryo sac is nourished by the nucellus, and the embryo sac further divides by mitosis into eight nuclei. This group of half-cells is now of the female gametophyte generation. One of these nuclei is the egg nucleus, another a fusion nucleus.

Fertilisation of the embryo

Shortly after formation of the pollen grains, the cells in the anther wall dry up and shrink so that the pollen sacs burst open when the anther splits. The pollen grains are released and travel to a receptive stigma.

Receptive stigmas will be sticky, catching any passing pollen, whether it be carried by the wind or a visiting animal. Interactions between proteins on the coat of the pollen grain and the stigma enable pollen grains to 'recognise' their compatibility. If compatible, then a pollen tube will rapidly grow down through the style and into an ovary.

Growth of the pollen tube is controlled by the pollen tube nucleus, and secretions of digestive enzymes enable it to penetrate into the style. At this time, the generative nucleus of the pollen grain divides by mitosis to form two male gametes. Ultimately, the pollen tube will enter the embryo sac. The pollen tube nucleus degenerates and the two male gametes enter the embryo sac through the pollen tube.

One male gamete will fuse with the egg cell, and the other with the fusion nucleus to form an endosperm nucleus (which will eventually become the food store of the seed – the endosperm, see p. 75). Thus the process of fertilisation in flowering plants is actually a double fertilisation, and the result is a rudimentary embryo (see chapter 5) and endosperm.

Apomixis – seeds without sex

Some types of plant are capable of producing viable seeds without fertilisation. This is called apomixis, and it combines the characteristics of vegetative reproduction with the opportunities created by seed dispersal. Some species of *Sorbus* are apomictic, as are dandelions (*Taraxacum*). Apomixis can lead to the evolution of small, isolated populations in the wild. It also means gardeners can raise certain species from seed without variation.

Sorbus intermedia,
Swedish whitebeam

Franz Andreas Bauer
(1758–1840)
Ferdinand Lukas Bauer
(1760–1826)

The Bauer brothers, Franz Andreas and Ferdinand Lukas, were accomplished botanical artists, although Franz was less well known than his much-travelled and renowned brother. They were born in Feldsberg, Moravia, now Valtice in the Czech Republic, Franz in 1758 and Ferdinand in 1760. Their father was the court painter to the Prince of Liechstenstein, and consequently they were surrounded by art and paintings from an early age.

Their father died a year after Ferdinand was born, and the brothers (including the eldest, Joseph Anton) were placed in the custody of Father Norbert Boccius, Abbot of Feldsberg, who was a physician and a botanist. Together, while still in their teens, the three brothers recorded all the plants and flowers in the monastery garden and produced more than 2,700 watercolour paintings of plant specimens. Their work illustrated a remarkable 14-volume book, *Liber Regni Vegetabilis* (Book of the Plant Kingdom) or the *Codex Liechtenstein*. Under Boccius' guidance, Ferdinand in particular became an astute observer and lover of nature.

In 1780, Franz and Ferdinand went to Vienna to work for Baron Nikolaus Joseph von Jacquin, who was a renowned botanist and artist, Director of the Royal Botanical Garden at Schönbrunn Palace and Professor of Botany and Chemistry at the University of Vienna. There they were introduced to the Linnaean taxonomic system and the use of microscopy to record fine detail, perfected their skills as botanical illustrators and concentrated on exact observation of plants. They developed an extraordinary attention to detail that later became their hallmark. The careers of the two brothers then diversified.

In 1788, Franz travelled to England, became known as Francis, and settled at Kew where he spent the rest of his life. He worked at the Royal Botanic Gardens, Kew, for more than 40 years and was given the title Botanick Painter to His Majesty, under the sponsorship of Sir Joseph Hooker. His illustrations were a valuable scientific record of the newly discovered plants from around the world that were introduced to Kew, where they were grown and studied for the first time in a scientific manner. Unlike his brother, he did not want to travel and became more interested in the science and botany of the plants he studied.

Franz was elected a Fellow of the Linnean Society and became a Fellow of the Royal Society. He died at Kew in 1840.

Franz Andreas Bauer (above) and his brother Ferdinand spent their lives illustrating plants for books and collections.

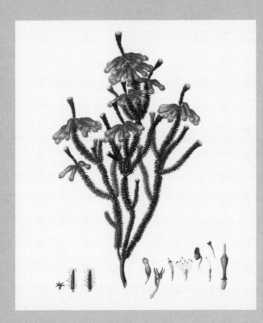

Erica massonii,
Masson's heath

This illustration by Franz Andreas Bauer was used in *Delineations of Exotick Plants* cultivated in the Royal Garden at Kew. It shows botanical characteristics displayed according to the Linnean system.

Ferdinand went on to become the more famous botanical artist of the two. He travelled with botanists and explorers to record plants in their natural habitats and in the context of the local natural history.

In 1784, Ferdinand accompanied John Sibthorpe, a botanist and professor from Oxford University, on his travels to Greece. This resulted in the publication of *Flora Graeca*, which is illustrated throughout with Bauer's magnificent artworks of the Greek flora.

He then travelled to Australia on HMS *Investigator*, accompanying Sir Joseph Banks' botanist, Robert Brown, as botanical draughtsman on Banks' recommendation. Bauer made around 1,300 drawings of the plants and animals that were seen and collected during the voyage. His coloured artworks revealed the wonders of the Australian flora and fauna and some of the paintings were published as engravings in *Illustrationes Florae Novae Hollandiae*. This was the first detailed account of the natural history of the Australian continent.

When HMS *Investigator* set sail to return to Britain, Bauer remained in Sydney and took part in further expeditions to New South Wales and to Norfolk Island.

He returned to Austria in 1814 and continued to work for English publications including Aylmer Bourke Lambert's *A Description of the Genus Pinus* and illustrations of *Digitalis* for John Lindley. He lived near the Schönbrunn Botanical Garden and spent his time painting and making excursions into the Austrian Alps. He died in 1826.

Several Australian plant species were named after Ferdinand Bauer, and the genus *Bauera* and Cape Bauer on the Australian coast were named after him.

The standard author abbreviation F. L. Bauer is used to indicate him as the author when citing botanical names of plants.

Thapsia garganica,
Spanish turpeth root

One of Ferdinand Bauer's many magnificent artworks of Greek flora used to illustrate the book *Flora Graeca*.

Plant breeding – evolution in cultivation

The mechanisms of plant evolution are the same in nature as they are in cultivation, the only difference being the pressures of selection. In nature, for example, plants that show weak traits are soon killed off – a process known as natural selection or 'survival of the fittest'. In cultivation, plant breeders only select those plants that show desirable characteristics (such as large flowers, higher yields or colourful leaves) – the rest will be discarded in a process known as artificial selection.

Plant selection

The origins of plant breeding can be traced back to the dawn of agriculture about 10,000 years ago, when hunter-gatherers collected crops from wild sources.

At this time, humans were nomadic, following animals (their main source of food) and gathering food from plants as they moved around. Dogs, pigs and sheep were the first animals to be usefully controlled and domesticated, which led in turn to a more semi-nomadic lifestyle. Early farmers found that some of the plants that were useful to them could be stored or regrown from seed.

As time moved on and farmers learned from experience, the range of crops that was grown was extended, and farmers began to select only the best plants from the crop for growing next season. As a result, a slow but gradual process of crop improvement began to take place.

Thousands of years of refinement through cultivation now means that some crops vary greatly from their wild ancestors, such as the potato (*Solanum tuberosum*) and sweetcorn (*Zea mays*). The cultivated forms of these crops can only thrive in a man-made environment and could not survive in the wild. Some crops, like cocoa (*Theobroma cacao*) and leeks (*Allium porrum*), however, have not changed much from their wild cousins. Some, like sainfoin (*Onobrychis viciifolia*) – once a popular legume – have actually dropped out of mainstream cultivation.

Trees and shrubs take a very long time to reach maturity, and many nomadic or semi-nomadic tribes would not have been in one place long enough to make use of them as crops. Consequently, tree and shrub crops (mainly fruit) have only been introduced to cultivation relatively recently as humans became less transient. Settlements founded near natural apple and

Zea mays, sweetcorn

pear forests in western Asia led to the first orchards; as the forests were cleared for fuel and building materials, the most productive trees would have been left standing.

Much more recent is the dawn of ornamental agriculture, or horticulture as it is known. The first gardeners would have used the same techniques of selection, but rather than focusing on yield they would have been more interested in a plant's ornamental merits. Crop plants such as figs, olives and grapes would have been obvious choices for the first gardens, since they were already domesticated and produced a useful crop. Some of these may also have had some kind of religious or cultural significance, a factor that may have extended to a select few non-crop plants, such as roses, yews and laurel trees. Some plants may simply have been chosen because they were easy to grow or provided shade and shelter. Through plant breeding, by AD 1200 five groups of ornamental roses were already recognisable: Albas, Centifolias, Damasks, Gallicas and Scots.

Discovery of new worlds between the 15th and 18th centuries brought many new plants into cultivation. Some, like maize or sweetcorn (*Zea mays*) are now a transglobal commodity at the forefront of plant-breeding science, while others have become extremely dependable garden plants, such as the flowering currant (*Ribes sanguineum*).

The golden age of ornamental plant discovery must be the 19th century, when plant hunters were sent out across the world to seek new species, which were duly sent home, unpacked and propagated on. As the world became 'smaller' the work of plant hunters became more intense and localised. It continues through the travels of people such as Bleddyn and Sue Wynn-Jones of Crûg Farm Plants in Wales, who have introduced many new garden plants, particularly from Asia.

Vitis vinifera, grapevine

New discoveries often take time to become widely accepted by gardeners, but they do increase the material with which plant breeders can work. As an example, gardeners are beginning to see new types of hellebore appear in the marketplace, such as *Helleborus* × *belcheri*, a hybrid between the long-cultivated *H. niger* and the recently introduced *H. thibetanus*.

The science of plant breeding continues today at a fierce pace. In the world of agriculture we are very fortunate in having within easy grasp a whole multitude of different fruits and vegetables. In the world of gardening the latest *RHS Plant Finder* lists a staggering 75,000 ornamental plants in current cultivation – and those are just the ones available in the UK.

Modern plant breeding

Selection still plays a big part in plant breeding, but it can be a slow process, sometimes taking up to 10 years before a cultivar is released to gardeners. It generally involves three steps: selection, elimination and comparison.

Firstly, a number of potentially good plants are selected from a chosen population, which usually shows a large amount of variability. These chosen plants are grown for observational purposes, over several years and under different environmental conditions, and the poorest performers eliminated. Finally, the plants that remain are compared to existing varieties for improved performance.

Sometimes the process can be much quicker. Keen-eyed plant breeders may spot natural mutations or variations (see p. 121). These are then increased by vegetative propagation and subjected to observation and test. By this method gardeners have gained the Irish yew (*Taxus baccata* 'Fastigiata') with its distinctively upright shape, and the nectarine (*Prunus persica* var. *nectarina*) – which is essentially a hairless peach.

Hybridisation

With the aim of bringing together desirable traits from two different plants, hybridisation through cross-pollination is today a frequently used plant-breeding technique. The principles are based upon observations made in the 17th century, when the sexual function of the flower was revealed, yet it was not until the 19th century that plant breeders began to use this information in any practical way.

Before this time, 'primitive' hybridists would place two distinct cultivars in pots together when both were in full bloom, knowing that there was a reasonable chance that they would cross breed and produce seedlings bearing shared characteristics of both parents. The first repeat-flowering old roses were born this way, when the newly discovered *Rosa chinensis* was brought into cultivation in Europe and hybridised with the old roses already in gardens.

Rosa chinensis, China rose

Our modern understanding of plant heredity and genetics, first demonstrated by Gregor Mendel (see pp. 16–17) in the second half of the 19th century, has led to huge improvements in hybridisation techniques. The world of horticulture is populated by a good number of famous hybridists, such as Elizabeth Strangman – who introduced the first double-flowered hellebores – and Toichi Itoh, the Japanese breeder who first successfully crossed tree and herbaceous peonies to produce what are known in the peony world as Itoh (or intersectional) hybrids.

In the agricultural world, modern hybridisation techniques culminated in the 'Green Revolution' of the mid-20th century. Through the development of higher-yielding crop varieties along with modern pesticides, fertilisers and management techniques, the 'revolution' is credited with saving billions of people from starvation as the world's population grew to unprecedented levels.

F1 hybrids in plant breeding

The first step that a hybridist needs to take is to ensure that each parent plant is as 'pure' as possible, showing a minimum of genetic variation. This is usually achieved by self-pollinating the plants over a few generations so that they are inbred.

Once two pure lines have been generated, the second step is to allow them to cross-pollinate. The resulting progeny is then selected for combinations of the desired traits. Undesirable traits can be removed by repeated back-crossing with either of the parents.

Pollination under such circumstances must be controlled. Left to pollinate outdoors (open pollination), as would happen naturally, the pollen source could not be guaranteed – could be carried in from anywhere. The result would be widely varied progeny.

One of the biggest challenges facing hybridists, therefore, is avoiding introduction of pollen from other strains. If it is known that the pollen of a particular plant only travels short distances, then

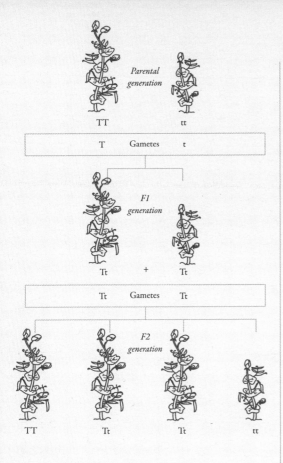

Enhanced breeding

With increasing knowledge and improved technology, breeders have developed ways to improve the speed, accuracy and scope of the breeding process. These include maintaining parallel selection programmes in northern and southern hemispheres, which allows two generations to be produced each year, and cultivating the plants in artificial growth rooms.

More recent laboratory techniques allow breeders to operate at the level of individual cells and their chromosomes. Genetic modification and genetic engineering allow scientists to insert new genes into existing plants to increase their resistance to disease, improve yield, or simply make plants resistant to certain herbicides so that weed control is made simpler.

To date, there is not much incentive to produce new garden plants through genetic modification. Work is focused on crop plants of international significance such as maize (*Zea mays*), tomato (*Solanum lycopersicum*) and wheat (*Triticum aestivum*).

Mutations and sports

Some plants produce natural mutations or variations that plant breeders may use as a basis for creating new cultivars. New plants created from spontaneous mutant or aberrant growth are known as 'sports', and such material may appear naturally (as in the nectarine) or it can be induced artificially, usually by exposure to chemicals or radiation.

In the 1950s, a form of peppermint (*Mentha* × *piperita*) that showed resistance to *Verticillium* wilt was produced through exposure to radiation, although it is no longer commercially available. Many modern cultivars of day lily (*Hemerocallis*) have come about as a result of breeding using chemicals to double their chromosome number. This results in cultivars with four sets of chromosomes (tetraploid), which usually have larger and more substantial flowers.

the fields that the parent plants are grown in can be isolated. Alternatively, pollination can be controlled by carrying out the crosses inside a glasshouse or polytunnel, or on a smaller scale by hand-pollinating flowers enclosed within a plastic bag or dome. This is known as closed pollination.

The first generation that results from cross-breeding is known as the F1 generation. The offspring are commonly referred to as F1 hybrids. F1 stands for 'filial 1', the first filial generation. F1 hybrids from carefully controlled cross-pollination will possess new and distinctive characteristics. The work involved in developing them costs a lot of money and takes several years, and as a result F1 seeds are more expensive than open-pollinated ones. Their one disadvantage is that they do not breed true – the seed they produce does not possess the same level of uniformity. Their seed, therefore, is not usually worth saving and sowing.

Syringa vulgaris,
common lilac

Chapter 5

The Beginning of Life

Growing plants from seed is one of the most satisfying aspects of gardening – seeing the seeds germinate and the seedlings develop into mature plants.

For plants, seeds are an essential part of their life cycle, the end result of sexual reproduction, ensuring the species lives on from generation to generation and providing a safety mechanism during unfavourable environmental conditions. For the gardener, they provide an inexpensive way of producing a lot of plants, especially annual bedding plants and many herbaceous perennials. All annual vegetable crops are grown from seed too, and you can also save seed from your own plants, providing a source of free seeds. For more information on seeds, see pp. 74–78 and pp. 110–115.

Development of the seed and fruit

Once fertilisation has occurred (see p. 115), the ovary is referred to as the fruit and the ovule is referred to as the seed. These are the two structures that will develop now that the embryo has begun to form. The seed consists of a seed coat enclosing an embryo (young plant) and endosperm (food store for the young plant). The embryo begins to expand by cell division (mitosis), and as it matures the first shoot (plumule), the first root (radicle) and the seed leaves (cotyledons) start to form.

As well as becoming the first leaves of the seedling, cotyledons may also have a food storage role, in addition to the endosperm (see p. 75). Monocotyledenous plants have just one seed leaf; dicotyledenous plants have two (see p. 28).

Ricinus communis, castor-oil plant

A selection of grasses

The endosperm nucleus undergoes repeated mitotic divisions to form the endosperm, which is a mass of cells, separated by thin cell walls, acting as a food store. The food is mostly stored in the form of starch, but oils and proteins can also form major constituents depending on the species; seeds of the castor-oil plant (*Ricinus communis*), for example, are high in oils, whereas those of wheat (*Triticum*) are high in both protein and starch. It is the endosperm of wheat that is ground into bread flour.

As the seed matures, the seed coat ripens and what was the ovary develops into a mature fruit. In berries and drupes, the ovary wall develops into a fleshy pericarp that serves to protect the seeds and aid their dispersal. One final development of the seed is the reduction in water content from about 90% by mass to just 10–15%. This reduces its metabolic rate considerably, and it is an essential step in seed dormancy.

Seed dormancy

A set of changes takes place in the seeds of most plants during ripening to ensure that premature germination cannot occur. Known as dormancy, it is a survival method that allows time for seed dispersal and promotes germination during optimal environmental conditions. For example, many seeds produced in the summer or autumn would probably not survive if they germinated before the onset of winter, so mechanisms exist that ensure germination is only synchronised with the onset of spring.

Some seeds can remain dormant for very long periods indeed. Others cannot. Gardeners may be perplexed to find 'use by' dates on their seed packets, but these dates reflect the life expectancy of the seed. After this time, some of the seed embryos may no longer be viable. Many gardeners may also be aware of the adage 'one year's seeds; seven years' weeds' – this is in reference to how seeds of many plants simply lie dormant in the soil, only awakening when the soil is disturbed. The seeds of *Silene stenophylla* (narrow-leaved campion) can get buried in permafrost, and excavated seeds estimated to be more than 31,000 years old have been successfully germinated.

Dormancy is caused either by conditions within the embryo, or outside the embryo. A combination of factors is not unusual, as in many iris seeds, which show a mixture of physiological and mechanical dormancy.

Physiological dormancy

Physiological dormancy prevents germination until chemical changes occur within the embryo. Sometimes chemical inhibitors, such as abscisic acid (see p. 99), retard embryo growth so that it is not strong enough to break through the seed coat. Some seeds exhibit thermodormancy, being sensitive to either heat or cold; others show photodormancy or light sensitivity.

Morphological dormancy

Morphological dormancy is seen where the embryo has yet to mature at the time of seed dispersal. Germination will not occur until the embryo has fully developed, which will cause a delay, sometimes further influenced by the availability of water or environmental temperatures.

Lathyrus odoratus, sweet pea

Physical dormancy

Physical dormancy occurs when seeds are impermeable to water or the exchange of gases. Legumes are typical examples as they have very low moisture content and are prevented from taking up water by the seed coat. Cutting or chipping the seed coat allows the intake of water, a practice recommended in the germination of sweet peas (*Lathyrus odoratus*).

Mechanical and chemical dormancy

Mechanical dormancy occurs when seed coats or other coverings are too hard to allow the embryo to expand during germination. Chemical dormancy relies on growth regulators and other chemicals that are present in the coverings around the embryo. They are washed out of the seeds by rainwater or by melting snow. Gardeners can simulate these conditions by washing or soaking the seed.

Seed germination

Germination is defined as the growth of a seed from the moment its embryo is triggered to grow (usually after dormancy) to the formation of the first leaves. Three basic conditions must be met before germination can take place: the embryo must be viable (alive), any dormancy must be broken (see p. 125) and the correct environmental conditions must be present.

Environmental conditions, however, can soon turn unfavourable. If all the seeds have germinated then potentially they could all be killed at this vulnerable stage in their life by an unfortunate turn in the weather. A clever adaptation seen in many plants, known as staggered germination, acts as an insurance policy; those that do not germinate in the first instance are delayed and germinate a bit later.

Breaking dormancy

Dormant seeds must have their dormancy 'broken' before germination can take place. Common triggers include high or fluctuating temperatures, freezing and thawing, fire or smoke, drought or exposure to the digestive juices of animals. Seed dormancy may be virtually non-existent in some cultivated plants, having been deliberately bred out of them.

Stratification

Gardeners must sometimes mimic nature to get their seeds to germinate when they want them to, by artificially breaking dormancy. There are a number of methods. Stratification is the first, and requires the gardener to give the seeds a bit of a nudge in the right direction. *Echinacea* seeds, for example, are unreliable germinators, but they can be encouraged to do the right thing if put in a refrigerator for a month. Such treatments work well for many seeds of deciduous trees and shrubs. Some seeds need a period of warmth followed by a period of cold, and then another period of warmth; a heated propagator, as well as a refrigerator, is handy for such seeds.

Scarification

In some cases, a bit of muscle is required to break through a hard seed coat – a process known as scarification. In nature, this might occur via natural decay or by the actions of an animal. For the gardener who cannot afford to wait, the seed coat can either be rubbed with a file or (with small seeds) placed and shaken in a screw-lid jar lined with sandpaper – *Acacia* seeds respond well to this treatment. Cutting, chipping or pin-pricking the seed coat are other options.

Seeds of acacias, such as *Acacia catechu* (catechu tree) shown here, should be soaked in warm water for four hours before sowing, or scarified using sandpaper.

Soaking

Soaking in water can remove natural chemical inhibitors from the seed and allow it to take up water. The seeds are usually covered in hot water for 24 hours or until they visibly swell, and any floating seeds should be discarded. They must be sown right away. Sweet peas (*Lathyrus odoratus*) benefit from light scarification as well as soaking prior to germination.

In Australian and South African plants, fire and smoke are well-known triggers for breaking seed dormancy. High temperatures are sometimes enough to break dormancy, but it may first be necessary to physically release the seeds from their 'pods' (such as the gum nuts of *Eucalyptus* or the dry follicles of *Banksia*). It may actually be the chemicals in wood smoke that serve to break dormancy.

Germination factors

The three vital conditions necessary for germination of all seeds are water, the correct temperature and oxygen. In many cases, the presence or absence of light is also a key factor, as in the fine seeds of foxgloves (*Digitalis purpurea*), which must be sown on the surface of the compost as they need light to germinate.

Water

Water is vital as seeds have a reduced water content. For the cells within to become turgid again, water is imbibed (soaked up) through the micropyle (see p. 75). Water is also used to activate the enzymes that are needed to digest the food reserves in the endosperm. Water causes the seed to swell and the seed coat to split.

In some seeds, once water is taken in, the germination process cannot be stopped, and subsequent drying out is fatal. Others, however, can imbibe and lose water a few times without ill effect.

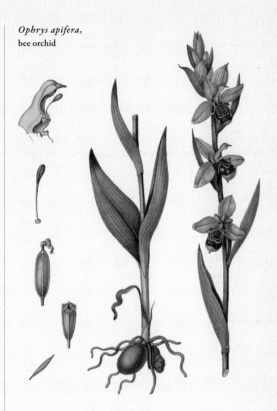

Ophrys apifera, bee orchid

Oxygen

Oxygen is required for aerobic respiration, allowing the cells to metabolise and burn energy. This will be the only source of energy until the seedling produces its first green leaves and starts being able to photosynthesise. Orchid seeds, which have no endosperm, must form a mycorrhizal relationship with soil-borne fungi upon germination, as it is the fungal partner that will supply the plant with its fuel for respiration.

Temperature

There is usually a characteristic temperature range outside which seeds will not germinate. Temperature influences the metabolic rate of the cells and the rate of enzyme activity. Too cold or too hot and the whole process will shut down – it is important for gardeners to know the requirements of their seeds and to maintain a stable temperature.

Light

In some seeds, germination is light-dependent. This is useful if the seed is buried, as it will only germinate if it is brought close to the surface. Buried seeds might not have sufficient food reserves for their shoots to reach the surface upon germination. Most seeds are not affected by light levels, but those that are will not germinate unless there is sufficient light for photosynthesis; such seeds contain a light-sensitive pigment called phytochrome. The mechanism is often seen in woodland species and is probably an adaptation that allows plants to germinate only if there is sufficient light for growth, which may come about when a large tree falls down. Foxgloves (*Digitalis*) are an example of flowers that might be seen in a forest clearing.

***Digitalis lutea*,**
small yellow foxglove

The seeds of this woodland plant need light to germinate, so do not cover or bury them when sowing – just press lightly into the compost.

The physiology of germination

A typical seed stores reserves of carbohydrates, fats, proteins and oils in the endosperm and in the cotyledons and the embryo. The main reserves are oils and starch (a type of carbohydrate). As a result of the imbibition of water, the embryo becomes hydrated, which activates enzymes to kick-start the whole germination process.

The two centres of activity are the endosperm and the embryo, and the enzyme-driven reactions are either catabolic (where larger molecules are broken down into smaller units) or anabolic (where smaller units are built into larger units). Catabolism of the food reserves can be summarised as proteins being broken down into amino acids, carbohydrates into simple sugars (such as starch into maltose and then into glucose) and fats into fatty acids and glycerol.

These smaller units then go through a series of anabolic reactions to construct new cells as the embryo starts to grow. Amino acids are assembled into new proteins, glucose is used to construct cellulose, and fatty acids and glycerol are used to construct cell membranes. Plant hormones are also synthesised, which influence the germination process. Glucose is used as a fuel – it is translocated to the growth regions of the embryo to help cell metabolism.

A net loss in the dry weight of the seed occurs during the first few hours, as the seed uses up its food reserves. It is not yet able to photosynthesise and manufacture its own food. This weight loss continues until the seed produces its first green leaves, and the endosperm begins to shrivel and wither.

Growth of the embryo

In the embryo, three stages of growth are manifest: cell division, enlargement and differentiation.

Cell division

The first visible sign of growth is the emergence of the embryonic root, called the radicle. The radicle is positively geotrophic, meaning it grows downwards and anchors the seed in the ground. Cell division and expansion occurs in an area at the base of the radicle called the epicotyl. The radicle is covered in fine root hairs, and these begin to absorb water and minerals from the soil.

Enlargement

The embryonic shoot, called the plumule, is negatively geotrophic, and it grows away from the pull of gravity, in an upwards direction. The growth centre of the plumule is called the hypocotyl, and like the epicotyl it is found close to the seed, not at the tip of the plumule.

Differentiation

There are two ways in which the seedling can extend its first shoot above the soil into the air. Either the radicle grows, forcing the seed out of the soil, or the plumule grows, leaving the seed below ground. The former is epigeal germination, and the seed comes out of the ground with its cotyledons (seed leaves) greening and opening. Sometimes the seed coat remains attached to one of the seed leaves, a clear sign of epigeal germination. Courgettes and pumpkins are typical examples.

If the plumule grows, then hypogeal germination takes place. The cotyledons are forced to stay underground as the plumule elongates and forms the first leaves. The cotyledons wither and decompose. Peas are an example.

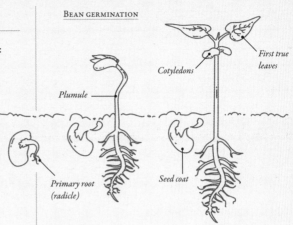

BEAN GERMINATION

In epigeal germination, the plumule elongates and pulls the cotyledons and young stem upwards through the soil.

In monocotyledonous plants, such as grasses, the emerging root and shoot are covered by protective sheaths, called the coleorhiza and coleoptile, respectively. Once the coleoptile emerges from undergound, its growth ceases. The first true leaves then emerge from this protective sheath.

The emergence of the seedling

With the emergence of the radicle and plumule, the seed has entered its seedling stage. Germination is complete, and the road to maturity begins. This is a vulnerable stage for all plants as seedlings are prone to damage by herbivores and pests, susceptible to diseases and maybe low or high temperatures, waterlogging or drought.

Many plants simply produce as many seeds as possible in the hope that a few at least will establish successfully. Some other plants adopt the opposite approach, by investing all their energies into just a few seeds. In these cases, the rate of germination and successful establishment must be relatively high if the plants are to succeed. Often the services of an animal are employed, and some kind of reward, perhaps in the form of a fleshy fruit, is offered (see pp. 78–81).

Matilda Smith
1854–1926

Matilda Smith was a botanical illustrator who was born in Bombay (Mumbai) and came to England in her infancy. She is most famous for her 45-year career supplying fabulous plant illustrations for *Curtis's Botanical Magazine*.

The first issue of the magazine, then titled *The Botanical Magazine*, was printed in 1787. Now well over 200 years old, the magazine is the longest-running botanical periodical featuring colour illustrations of plants. Each four-part volume contains 24 plant portraits reproduced from watercolour originals by leading international botanical artists. From 1984 to 1994 the magazine appeared under the title of *The Kew Magazine*, but in 1995 it returned to its roots and the historical and popular name *Curtis's Botanical Magazine* was reinstated.

Sir William Jackson Hooker was the first Director of the Royal Botanic Gardens, Kew and the magazine's editor from 1826, bringing his wealth of experience as a botanist. Joseph Dalton Hooker followed on from his father, becoming Director of Kew Gardens in 1865, and consequently editor of its magazine. During this time, Walter Fitch withdrew his services as botanical illustrator to Kew, having been its principal artist for 40 years. The loss of Fitch meant that the survival of the magazine depended on Joseph Hooker recruiting and training a new, dedicated illustrator to continue this vital feature.

Matilda Smith was a prolific plant illustrator for Curtis's *Botanical Magazine*.

Rhododendron concinnum, rhododendron

This illustration of *Rhododendron concinnum* by John Nugent Fitch is from an original by Matilda Smith, published in *Curtis's Botanical Magazine*.

Hooker, who was a botanical draughtsman of considerable ability himself, knew his second cousin Matilda Smith had artistic talent and decided to train her further and supervise her work. Within a year, Matilda's first illustration appeared in the magazine. From 1878 to 1923, Matilda drew more than 2,300 plates to illustrate the pages of *Curtis's Botanical Magazine*.

She also illustrated many other publications, including more than 1,500 plates for Hooker's *Icones Plantarum*, which illustrated and described plants selected from the Kew Herbarium. She was also responsible for reproducing missing drawings for rare, but incomplete volumes in the Kew Library, and was said to have produced more coloured drawings of living species than any contemporary artist.

Pandanus furcatus,
Himalayan screw pine

This illustration was drawn by Matilda Smith and Walter Fitch for *Curtis's Botanical Magazine*.

After 20 years of steady output, her exceptional skills and contribution to the magazine earned her official admittance to the Herbarium staff as the first official botanic artist of Kew, and hence the Civil Service's first ever botanical artist.

Matilda Smith was also noted for her skill in reanimating dried and flattened specimens, often of an imperfect character. She illustrated a number of other books, including *The Wild and Cultivated Cotton Plants of the World*, and was the first botanical artist to comprehensively illustrate the flora of New Zealand, in a book written by Joseph Dalton Hooker.

Because of her enormous contribution to botanical illustration, Matilda was made an Associate of the Linnean Society – the second woman to have achieved this. She was also awarded the Royal Horticultural Society Silver Veitch Memorial Medal for 'her botanical draughtsmanship, especially in connection with the Botanical Magazine'. And she was the first woman to be appointed President of the Kew Guild, an organisation of senior employees of Kew Gardens.

The plant genera *Smithiantha* and *Smithiella* were named in her honour.

The abbreviation M. Sm. is used to indicate her as the author when citing botanical names of plants.

Rhododendron wightii,
Wight's rhododendron

One of Matilda Smith's many rhododendron illustrations used in *Curtis's Botanical Magazine*.

Sowing and saving seeds

We have discussed natural methods of seed dispersal on pp. 78–80, but the seeds of cultivated plants have entered another realm of dispersal altogether: that of an intimate relationship with humankind. Since the dawn of agriculture, humans have collected and saved seeds, and this process continues today. They are traded and sometimes shared around the world, and in many cases seeds, or the fruit in which they are contained, are a valuable global commodity (such as wheat grains or cocoa beans).

Quercus suber,
cork oak

If they are not eaten or processed into something else, some of these seeds will arrive back in the hands of the gardener or farmer, ready to be sown again. It is an unusual life strategy, but every time a seed is sown, the job of its dispersal is complete.

Handling seeds and seedlings

Seeds derived from a commercial source may have been pre-treated in some way to protect them or make them easier to handle. First-time growers of sweetcorn may be surprised to see that the seeds are actually shrivelled grains of corn, sometimes treated with a dye to discourage human consumption. At other times the seeds will have been treated so that they germinate quickly. Pelleted seeds are embedded in a pellet, which makes them easier to handle, and sometimes seeds are coated with a fungicide. Seeds sometimes also come on water-soluble tapes or mats, making things really easy for the gardener.

Gardeners usually sow their seeds by hand, either in trays or pots, into a seed bed, or directly where they are to grow. The specific needs of plants vary hugely, but fortunately advice is usually at hand on the back of a seed packet. Sowing instructions often reflect the way a plant has evolved in its natural habitat, with the seeds of foxgloves (*Digitalis purpurea*) requiring light and the acorns of many oaks (*Quercus* spp.) needing to be sown deeply (as they are often collected and buried by animals).

Experienced gardeners often have their own tricks and tips, often passed down through the generations. Sometimes seed can be pre-germinated on a layer of cornflour gel or between two sheets of kitchen towel. Many vegetable seeds can be encouraged to germinate in this way and it is a good way (in cool climates) of getting a head start on the growing season and to ensure that you only sow viable seeds.

Seeds that can be sown directly where they are to be grown can be scattered or broadcast, or they can be sown in drills or furrows. This is a much more earthy and appealing way to sow seeds, but unfortunately it does not work for all seeds, particularly those vulnerable to pests at the seedling stage. The main secret to success is to know when to sow directly and when not to, and to prepare a good seed bed, free of weeds and large stones, with a fine and crumbly structure.

For the gardener, nothing could be easier than when a plant sows itself. The resulting seedlings can then simply be carefully lifted out of the ground and potted on until they are larger and can be replanted elsewhere. Self-seedlings, however, often do not come

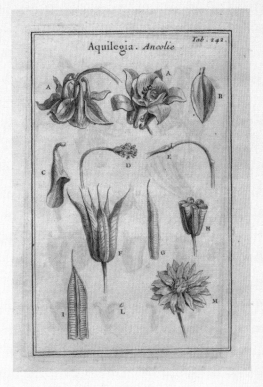

A plate of *Aquilegia* (columbine, granny's bonnet), drawn by Joseph Pitton de Tournefort, showing the seed heads.

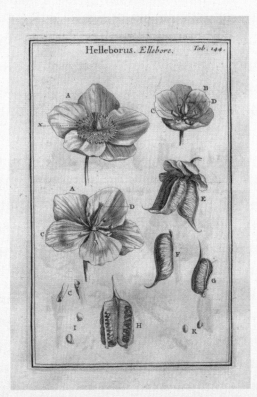

A plate of *Helleborus* (hellebore), drawn by Joseph Pitton de Tournefort, showing the seeds and seed heads.

true to type, the very characteristics that made the parents attractive become lost or diluted in the offspring. Hellebores and columbines (*Aquilegia*) are classic examples; a pity, because both can self-seed freely. Some cultivars, however, do come true from seed, such as the wonderful woodland ground cover plant known as Bowles's golden grass (*Milium effusum* 'Aureum').

On a commercial scale, seeds are usually sown by machine in fields, or they are grown in massive seedbeds and then transplanted. The latter is especially true of tree seeds. The cost of individual handling of seeds and seedlings is often prohibitive, so herbicides, pesticides, fungicides and other agrochemicals are employed to ensure the seeds and seedlings establish well.

Seed saving

When we discuss seed saving we enter the world of ethnobotany – the study of the relationship between plants and people. Often there are particular plants that have been selected to grow well in specific localities, and these may carry a great deal of cultural significance, such as the quinoa plant (*Chenopodium quinoa*), which was sacred to the Incas.

Throughout the history of mankind, the seeds of cultivated plants have passed through millions of pairs of hands over thousands of generations. As a result, the diversity of our crop and ornamental plants is extensive. Saving seed is a human endeavour, and preserving it, and adding to its diversity, is probably essential to the future of the human race.

On an individual level, however, saving seeds is all about collecting seeds for the next crop. Maybe only the seeds from the best plants are saved, perhaps ensuring a degree of improvement. It is only through the collected actions of our ancestors that we have the seed heritage that we have now.

The work of seed libraries and seed banks across the world, therefore, is not a folly, but a sensible step forward. Pioneers such as Nikolai Vavilov in the mid-20th century paved the way for contemporary seed banks. Now we have many not-for-profit and government-sponsored seed-banking organisations, of various sizes and ambitions, across the world. These include the Global Crop Diversity Trust in Germany, the Peliti Seed Bank in Greece and Seeds of Diversity in Canada.

The most ambitious projects to date are Kew's Millenium Seed Bank in the UK (with approximately fifty partner countries worldwide), the 'Doomsday Vault' in Norway (Svalbard Global Seed Vault), which is built to withstand global catastrophe, and the UPM Seed Bank in Spain – devoted to wild species of the western Mediterranean.

While gardeners can save seed from just about any plant, those from highly bred F1 hybrids are not usually worth saving, as they will not come true to type and the resulting seedlings will have a wide range of characteristics. Similarly, saving and sowing pips and stones from tree fruit cultivars will usually result in plants that produce few or poor quality fruit.

Seed should be collected once it is completely ripe, but before it is dispersed from the plant – sometimes a very fine line, so keep an eye on plants as the seedheads mature. Collecting seeds before they are fully ripe will result in poor germination, or they may deteriorate and rot in store. Whenever possible, collect the seed on a fine, dry day as this will help ensure the seed is perfectly dry. Seeds have to be dry for storage, since damp seed will rot or go off very quickly.

Chenopodium quinoa, quinoa

Quinoa is a grain-like crop whose edible seeds are an important staple in Andean cultures. The Incas held it as a sacred crop.

Seed banks in the soil

Nature has its own way of saving seeds, and that is in the soil. In any sample of soil, the seeds of hundreds of plants lie dormant, ready to germinate should conditions become favourable. The ecological significance of this is considerable, for it allows plant life to quickly re-establish after disturbance or catastrophe. Anyone who has witnessed the devastation caused by a forest fire, and how little time it has taken nature to reclaim the land, will not be able to deny its effectiveness. The famous poppy fields that established themselves on the abandoned battlefields of France and Belgium during the First World War in 1915 were the result of weed seeds germinating on the churned-up soil.

Unfortunately, gardeners will have little time for the bank of seeds that lies in their soil. Mostly weeds, these will present a gardener with a persistent headache, as he or she attempts to eradicate them. The old adage, 'One year's seeds, seven years' weeds' is a warning that if weeds are left unattended, and allowed to go to seed (which some of them do quite quickly and readily), then the seeds that lie dormant on the soil could remain there dormant for quite some time.

In truth, seeds can remain dormant from anything from just a few months to over 100 years. Although this sounds daunting, a well-cultivated garden or plot that has been carefully tended for many years will suffer much less from weeds than one that is neglected.

Needless to say, though, if any garden is left alone for a season or two, nature will soon reclaim what is rightfully hers, no matter how well tended in the past. It is a reminder that a garden is nothing more than an artificial construct, made to serve the purposes of man. The Lost Gardens of Heligan in Cornwall, UK, are a case in point, forgotten for many years yet painstakingly reclaimed from nature in the 1990s.

Papaver rhoeas, field poppy

BOTANY IN ACTION

Competition for life

In a garden, the gardener acts like a referee. Plants are allowed to grow here but not there, weeds are exterminated and unwanted or poorly performing plants are replaced with new ones. And yet in a tightly packed garden, plants still face a struggle for life. A garden is made up of a number of special habitats: walls for plants to climb up, shady spots for woodland perennials, a sunny spot for a tree. Viewed in this way, a gardener can help his plants in their quest for life. Choose the right plant for the right place, and then make sure each has what it needs for satisfactory growth. Good use of all available habitats will lead to a full and varied garden.

Tulipa,
tulip

Chapter 6
External Factors

The subject of this chapter concerns the external environment of plants. As plants are unable to move around, they have to endure whatever the environment throws at them, from hot sun and drought to extremely low temperatures, ice, snow and high rainfall. As a result, the external environment has a profound effect on all plants.

All plants have therefore evolved to tolerate most of the extremes that their natural habitat throws at them. For example, plants from the tropical rainforest have to be able to cope with high rainfall and humidity, poor and shallow soils, and either low or intense light levels depending on whether they are growing in the forest understorey or high up in the tree canopy. Desert plants are adapted to an extremely arid environment, having to make best use of the brief periods of intermittent rain, and plants from Mediterranean ecosystems inhabit a fire-prone environment with low summer and high winter rainfall and poor, well-drained soils.

The majority of plants we grow in our gardens originate from cool- or warm-temperate regions. Within such a range, the tolerance of different plants is highly varied, but most will have to endure variable summer temperatures, freezing winter temperatures, high winds and rainfall, as well as the occasional period of drought.

The soil

Soil is not just earth, it is a whole ecosystem. When it comes to external factors acting on a plant, it is perhaps the most critical zone of action. It is the only place where air, water, rock and living organisms all come together. As the old saying goes – 'The answer lies in the soil'.

One of the key functions of soil is to support life, and the soil is absolutely teeming with life, from microscopic organisms to larger insects and earthworms. It provides the anchorage medium for plants, ensuring they remain securely rooted and stable, as well as supplying the vast majority of the water and essential mineral nutrients needed for their growth. Its structure, make-up and content are critical, so much so that there is a whole area of science devoted to its study: soil science.

The perfect soil for gardening is easy to dig and work, warms up quickly in spring so plants start growing quickly and holds adequate amounts of water for healthy plant growth, yet is able to drain well so that it never becomes waterlogged. Good fertility is also important, with lots of essential plant nutrients, and plenty of organic matter that benefits soil structure as well as the flora and fauna that live there. Sadly, gardeners are often faced with something far away from this ideal and need to spend some time and effort on soil improvement.

Why are soils different?

Anyone who has gardened in more than one garden will be aware that soils can vary considerably from one area to another. Some are thick and heavy, others light and free-draining. There is an enormous variety, and their make-up and behaviour depend on their geological, topographical or human history.

In geological terms, the raw material of a soil is formed from the erosion and weathering of the underlying rocks into mixtures of different particle sizes. Soils are largely made up of these weathered solids. In topographical terms, the surface landscape, its climate and its exposure to the elements determines the patterns of soil erosion, deposition and drainage. It will also affect the rate and pattern of the build-up of organic waste.

On a human level, people interfere with natural soil types through cultivation and soil improvement. Draining may be improved, soil pH adjusted and nutrient content altered by the addition of fertilisers. Human activity can also have an extremely detrimental effect on soils by increasing the amount of erosion by

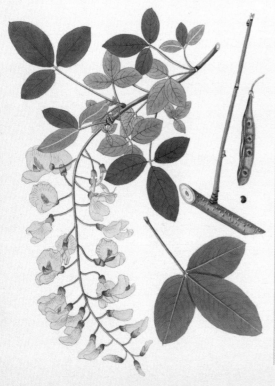

Laburnum anagyroides,
laburnum, golden chain tree

the removal of plants, as well as compaction, where all the air is squeezed out of a soil by heavy traffic. The soils of Mediterranean Europe are believed to have been progressively degraded by human mismanagement.

There are a number of ways in which the different soils can be described. Gardeners are probably most familiar with the terms 'soil structure' and 'soil texture'. Both are different measures of soil type, used together as they influence each other. Soil maps exist that show the diversity of soil types and properties in a given area.

The soil profile

If a hole is made of sufficient depth, then a vertical section of the soil can be viewed, known as a soil profile. From this, important information about a soil's structure and fertility can be gathered. The profile changes with depth, with the mineral origins of the soil becoming clearer the closer you get to the base rock. There are two main sections to the soil profile:

Topsoil

Topsoil is the upper layer of soil, where plants generally concentrate their roots and obtain most of their nutrients. It varies in depth, but in most gardens it is likely to be approximately the depth of one spade blade, i.e. about 15 cm (6 in) deep. The topsoil is the most fertile zone as it has the highest concentration of organic matter and microorganisms. Its depth can be measured from the surface to the first densely packed soil layer, the subsoil.

A soil profile will help a gardener to determine the depth of the topsoil, and it will also indicate how well the soil drains. The presence of any hard pans can be seen (hard layers that can form an impenetrable barrier to plants), as can the nature and size of the stones that may impede cultivation. Fine white roots throughout the topsoil indicate good drainage and aeration, and a dark or even black topsoil is a sign of plentiful organic matter, usually accompanied by plentiful earthworm activity. Worms aerate the soil and mix the organic matter. In very acid or water-logged soils, worms are absent.

Mineral matter in the topsoil is often red, orange or yellow. Blue and grey colours are not a good sign as they indicates poor drainage and little aeration, and such soils may smell foul. Aeration is not only important for the growth of plants, but also for the activity of bacteria, particularly those that fix atmospheric nitrogen. White mineral deposits are usually calcium carbonate – limestone or chalk.

Subsoil

Subsoil, like topsoil, is composed of a variable mixture of sand, clay and / or silt, but it is more compact with fewer air spaces. It also has a much lower percentage of organic matter and as a result it is usually quite a different colour to the topsoil. It may contain roots of deep-rooting plants, such as trees, but the majority of plant roots lie within the topsoil.

Below the subsoil is the substratum, the residual parent bedrock and its sediments. In some areas, geological action may have resulted in the deposition of soil derived from bedrock some considerable distance away, by the action of glaciers or rivers. The topsoil and subsoil may then show a different mineral composition to the bedrock. For example, a chalky soil may overlay a granite bedrock.

Gardeners must be careful not to mix subsoil with topsoil as it will have a detrimental effect on the soil, affecting its structure, fertility and biological activity. It can take years for the soil to recover. The mixing of topsoil with subsoil is easily done when gardens are landscaped, or when building work requires the digging out of foundations. The topsoil must be removed first and kept separate from the subsoil. On the chalk downs of southern England, some of the topsoils are so thin that they cannot be ploughed deeper than 7.5 cm (3 in). To go below this would bring up so much chalk that it could exercise an injurious effect for years.

Soil texture

'Texture' describes the relative proportions of mineral and rock particles in the soil. It is a test that is easily done at home, because the texture of a soil can be felt between the fingers: is it sandy, is it gritty or is it sticky?

The International System of Texture Grades, first proposed by the Swedish chemist Albert Atterberg in 1905, classifies rock particles by size. Those larger than 2 mm are classified as stones, those less than 2 mm but larger than 0.05 mm are classified as sand, those between 0.05 mm and 0.002 mm are termed silt, and those less than 0.002 mm are clay particles. The physical behaviour of the soil is determined by the three smaller particles: sand, silt and clay. It is the relative proportions of these that soil texture measures, with the dominating particles giving the soil its main characteristics. Generally speaking, the largest particles

Weigela is ideal for clay or silt soils and is suitable for all types of soil pH.

– sand and small stones (such as grit) – are responsible for aeration and drainage, and the microscopic clay particles are responsible for binding with water and plant nutrients.

> ### BOTANY IN ACTION
>
> Scoop up a handful of soil, moisten it slightly, roll it in your hands and attempt to make a ball. If a ball cannot be formed then the soil is sandy. If a ball can be formed, then try to roll it into a sausage, then into a ring. If you can make a ring, the soil has a high clay content. If the sausage is unstable, crumbly or cannot be formed into a ring, then the soil is a mixture of particle types; if it feels silky, it has a high silt content.
>
> All three soil types present the gardener with problems when present in extremes, but in the right proportions their qualities can complement each other. An even mixture of all three particle types is known as loam. Loamy soils are good for gardeners as they are fertile, drain well and are easily worked. Twelve soil texture classes are recognised (see diagram).

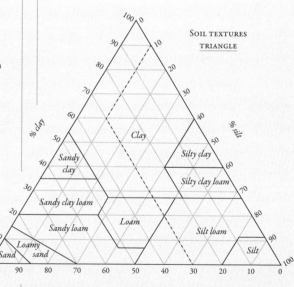

Sand
Loose when dry and not sticky when wet.

Loamy sand
With sufficient clay to give slight cohesion when moist.

Sandy loam
Rolls into a ball easily without being sticky but will collapse if pressed between thumb and forefinger.

Loam
Rolls into a ball easily and is slightly sticky, forming a crumbly sausage that cannot be bent.

Silty loam
As above but with a smoother, silkier texture.

Sandy clay loam
Can be moulded into a sausage that can be bent with careful support. It has a sticky feel, but still with the gritty texture of sand.

Clay loam
As above but less gritty.

Silty clay loam
As above but with a soapy feel.

Silt
With a distinct soapy or silky feel (pure silt soils are rare in gardens).

Sandy clay
Easily moulded into a sausage and bent into a ring. A gritty sand texture is evident.

Clay
Medium clay is as above but sticky, not gritty, giving a polished surface on rubbing. Heavy clay is very sticky and easily moulded into any shape. Fingerprints can be impressed into its surface.

Silty clay
Very sticky as above but with a distinct soapy feel.

Sandy soils

Known as light soils, sandy soils are easy to cultivate. They warm up much more quickly in spring than clay soils, but do not hold moisture well, so dry out very quickly. Sandy soils are also low in plant nutrients, since there is nothing to bind the nutrients and they are easily leached away by rain. Erosion is also a problem, because of their loose texture. Sandy soils are often acidic.

The moisture- and nutrient-holding capacity of sandy soils can be improved by adding liberal amounts

Clianthus puniceus (lobster claw) is an evergreen, scrambling shrub suitable for growing in mild areas.

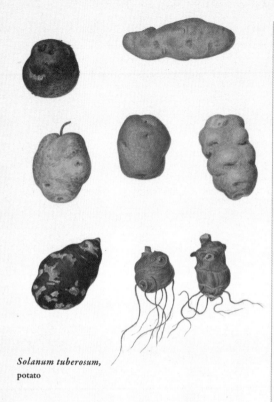

Solanum tuberosum, potato

of organic matter, which will bind the loose sand particles into more fertile crumbs and increase fertility. As they warm up quickly in spring, sandy soils are ideal for early crops like first early potatoes or strawberries, especially if they are on a slope facing the sun.

Clay soils

Known as heavy soils, the particles of clay are so small they are chemically active. They play an important role in holding mineral ions (nutrients) and in binding soil particles together, and it is for this reason that clay soils can be very fertile. Clay soils also hold a lot of water, owing to the capillary attraction of the tiny spaces between the particles.

When wet, clay soils are sticky, and they are also slow to drain. If poorly drained they may become very soggy or even waterlogged during wet weather; in this state they should not be worked or walked on, as this can lead to soil compaction. In summer, when the soil dries out, the clay particles tend to stick to each other rather than anything else and, as a result, the soil bakes hard and becomes impenetrable.

Although a heavy clay soil can be difficult to work, soils with a high clay component are some of the best gardening soils, as they have innate fertility and hold lots of nutrients. To improve a heavy clay soil, dig in lots of well-rotted organic matter and even sharp sand and grit. These will help break down the clay into smaller, separate crumbs, improving the overall structure and making the water and nutrients held within the clay particles more readily available to plant roots.

Silty soils

These are fertile, fairly well drained and hold more moisture and nutrients than sandy soils. On the downside, they are easily compacted and prone to erosion. Like other soil types, the answer to improving structure and fertility is by the regular addition of well-rotted organic matter, such as manure or garden compost.

Chalky soils

Chalky soils need to be considered separately. Chalk is not a soil type, but a rock or mineral. It is made up from billions of shells and skeletons of tiny sea creatures that were laid down when the land was covered by the sea, as in the southeast of England. Its presence in soils can make them very alkaline, and as a result oak trees hang on but don't thrive, birch and rhododendrons are virtually unknown and roses never really take off.

Chalk is fast-draining, which makes it fast to warm up in spring and quick to freeze, and easy to work. But plenty of organic matter needs to be added on a regular basis because it rots down quickly on chalk. Many gardeners underestimate how much organic matter they need to put into the soil each year, and no garden is as greedy for extra goodness as the one on chalk.

Some soils are completely dominated by organic matter, such as those on peat bogs or fenland where the dead remains of plants have not been able to rot down sufficiently, building up instead in the soil. It is from such soils that peat is extracted. One of the major limitations of the soil texture measurement system is that it takes no account of organic matter.

Soil structure

The structure of the soil depends on how well the textural components (or particles) are joined together. The presence of organic matter and clay causes the particles to clump together into 'crumbs', and connecting them is a network of pores, through which water, dissolved nutrients and air can circulate.

Soil structure influences the supply of resources to plant roots through water retention, nutrient supply, aeration, rain infiltration and drainage. Overall, it determines the soil's productivity. Approximately 60% of well-structured soil is made up of pore spaces; in poorly structured soils this can be as low as 20%. The action of roots, worms and microorganisms plays an important role in soil structure, as does the shrinking and expanding of the soil in hot and cold weather. Cultivation helps to improve soil structure.

The inclusion of well-rotted organic matter is one way to improve soil structure, particularly if one particle type (sand, silt or clay) is in abundance. It helps clay soils to 'flocculate' (form crumbs), as does the addition of lime and gypsum. The presence of calcium ions makes for good flocculation, as they are attracted to the negatively charged clay particles.

Soil structure is at its most vulnerable when wet, because the flocculants (the 'glue' that holds the crumbs together) become soluble. Just walking on wet soil can damage its structure, resulting in compaction and capping (where the top seals into a hard layer, causing puddling). For this reason, never cultivate a heavy soil in wet weather. Soils that are well drained are less prone to damage when wet, as there is less chance of the flocculants dissolving. Remember, if the soil sticks to your boots, it's too wet to dig.

There are four types of structural unit: platy, block, prismatic and crumb. The first three are found in the subsoil and are only of academic interest to the gardener. Crumb structure, however, is of high importance as it is manifest in the topsoil. Crumb is a rounded aggregate of small particles with well-defined pores. As its name suggests, it has a breadcrumb-like appearance, and it makes good topsoil. By raking, it can be made to form a good tilth – perfect for sowing seeds.

Blue plantain lily (*Hosta ventricosa*). Hostas need a fertile, moist but well-drained soil.

Soil pH

The term pH will be heard a great deal by gardeners. It is a measure of acidity and alkalinity and is expressed on a scale of 0 to 14, with a value of 0 being very acidic, 7 neutral, and 14 very alkaline. Most soils tend to range from 3.5 to 9. The optimum pH range for most plants is between 5.5 and 7.5. 'Chalky' and 'lime-loving' are terms associated with alkaline soils, and 'ericaceous' and 'lime-hating' are used in reference to acidic soils.

Soil pH is largely controlled by the parent rock and the mineral ions that leach from it. Magnesium and calcium ions are the most significant, and on soils rich in calcium, such as chalk or limestone, the pH tends to remain high (alkaline). Many soils are unable to fall below pH 4, and alkaline soils rarely exceed pH 8.

The acidity and alkalinity of a soil profoundly affects its behaviour, mainly because it controls the solubility of different minerals. This means that at different pH levels, some minerals are available to plant roots, and others are not. This is why some plants will grow on acidic soils (these plants are called calcifuges) and others will not (calcicoles).

Soil pH also has an affect on soil structure (through the availability of calcium ions), and on the activity of soil organisms that break down organic matter and recycle nutrients. While some plants are not too fussy about soil pH, some are very specific. Plants that only grow on acid soil are known as 'ericaceous', and they include rhododendrons and blueberries (*Vaccinium*). Some plants have a preference but not a requirement; for example, many fruit trees produce higher yields on a slightly acid soil (approximately pH 6.5).

The pH of a soil is literally set in stone, by its bedrock. It rarely fluctuates or changes. It is possible to influence the pH of a soil by incorporating mineral additives, such as sulphur to acidify a soil, or lime to increase the pH of a soil, but it can be expensive and over a relatively short time the pH will revert. The addition of organic materials with a high or low pH – such as spent mushroom compost (alkaline) or pine needles (acid) – often has little impact on soil pH. The best thing for gardeners to do if the plants they want to grow are not suited to the soil pH is to grow them in containers, using a potting compost that suits the requirements of the plant.

You can easily test the pH of your soil by buying a testing kit from your local garden centre. Take a soil sample a few inches deep from several points in the garden so that you get an average reading. If it is a big garden, there is a chance that soil pH may vary from one place to another.

Vaccinium uliginosum, bog bilberry

8.5	8.0	7.0	6.5	6.0	5.0	4.0
Moderately alkaline	Slightly alkaline	Neutral	Slightly acidic	Acidic	Very acidic	Very acidic

Soil fertility

The fertility of the soil has a direct influence on plants and how well they grow, since they absorb all their mineral nutrients from the soil through their roots.

Soil fertility is, in turn, dependent on its texture, structure and pH as described previously. It is not enough, therefore, for a gardener just to dose the soil with fertiliser and hope for the best. The soil must be cared for too – it is the cornerstone of good gardening.

Organic matter and humus

Humus is decomposed, stable organic matter in the soil. It is dark in colour and is just one of three types of organic matter in the soil. The others are fresh, undecomposed plant and animal litter, and those compounds formed from its breakdown. Humus is the result of the interaction of all these – having had more time to decompose.

Eventually, all organic matter is broken down into carbon dioxide and mineral salts by microorganisms. These mineral salts are important for plant nutrition, although most forms of organic matter contain relatively low levels of plant nutrients compared to fertilisers. Moderately undecomposed organic matter is an important foodstuff for soil organisms such as worms, slugs and snails. During its breakdown, some microbes produce sticky mucilaginous materials, which improve the crumb structure of the soil by holding soil particles together, and allowing greater aeration.

The presence of fulvic and humic acids in humus has a further effect on crumb structure as they bind to soil particles such as clay, changing their physical properties. A clay soil with added humus becomes less sticky and better aerated. Soils contaminated with heavy metals are sometimes treated with organic materials, as they can form strong chemical bonds with the heavy mineral ions, reducing their solubility.

Humus can hold up to 90% of its weight in moisture, and so its presence will help to increase a soil's water capacity and nutrient retention. Soils rich in humus will take on a much darker colour, which has a beneficial effect in spring as a dark soil will absorb more of the sun's energy and hence warm up more quickly.

The rate of humus breakdown is influenced by many factors. Conditions such as extreme acidity, waterlogging or nutrient shortages can inhibit the action of microorganisms, which leads to a build-up of surface litter and, in extreme and site-specific cases, peat. Such soils can be improved by the addition of lime, fertiliser and better drainage.

As a general rule, forest soils have the highest levels of organic matter, followed by grasslands and then farmland. Sandy soils have less organic matter than clay soils. An easy and effective way for gardeners to increase the organic content of their soil is by spreading any bulky material such as compost, leafmould and well-rotted manure. This can either be dug in or spread as a mulch, leaving the worms to take it into the soil.

Trillium erectum, **birthroot**

Trilliums naturally grow in woodland and need a fertile, humus-rich soil with plenty of added organic matter.

The nitrogen cycle

Although nitrogen is plentiful in the atmosphere, relatively few plants can absorb it directly as it needs to be in a form that can be taken up by the roots. It can either be supplied through the decay of organic matter, from the action of nitrogen-fixing bacteria, or by applying fertilisers.

The 'nitrogen cycle' describes the way in which nitrogen is converted from one chemical form to another. It takes place mostly in the soil, and bacteria play a key role. Soils with good structure will be at an advantage as plenty of air will be able to enter through the tiny pores in the soil and there will be a greater surface area for biological interaction.

The cycle begins with nitrogen from the atmosphere (in the form of N_2) being converted (or fixed) first into ammonia (NH_3), then into nitrites (NO_2) and then nitrates (NO_3) by soil bacteria. Nitrogenous waste from plant and animal detritus is also converted into nitrates by the same process. Nitrogen in nitrate or ammonium form can be assimilated by plant roots through the root hairs and used to build organic molecules.

NITROGEN CYCLE

The nitrogen cycle, which takes place mostly in the soil, describes how nitrogen is converted from one chemical form to another.

Other sources of ammonia, nitrites and nitrates come from lightning strikes, combustion of fossil fuels and from fertilisers. When plants die or are eaten by animals, fixed nitrogen will return to the soil and it may circulate many times there between organisms and the soil before it finally returns to the atmosphere in the form of N_2 by the action of denitrifying bacteria.

Nitrates are highly soluble and easily washed out of soils into streams and rivers. If soils that are heavily dosed with high-nitrogen fertilisers are allowed to leach into water courses it can cause considerable ecological problems, known as eutrophication. Algal blooms often result, as algae proliferate on the sudden upsurge in nutrients, leading to a serious reduction in the quality of the water. This in turn can lead to the death of many aquatic organisms, including fish and shellfish, as well as the land predators that depend on them for food.

Soil additives

While fertilisers are sometimes considered to be a soil additive, they are generally used to feed the plant rather than the soil. True soil additives include lime or farmyard manure, and they are used to improve the soil.

Lime

Lime is most important for its effect on the pH of the soil, and it is usually added to acidic soils to increase soil pH and in turn make more nutrients available. Gardeners must be aware that over-liming a soil can have the reverse effect, leading to nutrient deficiencies, so it is always best to assess the pH of a soil before any lime is added. There are, however, further uses for lime in the garden, such as its use on heavy clay soils to improve soil structure (see above).

Lime is also known to encourage the beneficial activities of earthworms and nitrogen-fixing bacteria, both of which dislike soils that are too acidic. It can

EXTERNAL FACTORS

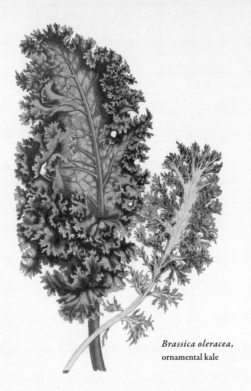

Brassica oleracea, ornamental kale

Bulky organic materials

These include a whole range of soil improvers, such as animal manures and home-made compost, although any green waste is usually OK, such as well-rotted compost from municipal green waste, straw, hay, mushroom compost and spent hops from a brewery. Remember that only *rotted* organic matter is added to the soil. Fresh material will need to be processed first, and this can be done by adding it to the compost heap for up to two years. Wood ash, in small amounts, is a useful additive to the compost heap as it is high in potassium and trace elements. As it also has a liming effect, it can be used in the same way as lime and applied to bare soil in winter.

Animal manures

Animal manures must be well rotted before they can be applied to the soil and are best used on the most greedy garden plants, such as tomatoes or roses. Making garden compost is an excellent way to dispose of garden waste, although many local authorities now pick up green waste alongside other household waste. Compost heaps are easy to make and many different designs of compost bin are available commercially. Other composting methods include bokashi and worm bins.

Leaf mould

Leaf mould is another bulky material that is easily made by stacking deciduous leaves and letting them rot down. It is a gradual business, taking up to three years, since they decay by the slow action of fungi. Unlike the bacterial activity in a well-aerated compost heap, fungi do not create a significant rise in the temperature of the leaf stack. Although leaf mould takes longer to make than garden compost, one advantage is that after initial stacking it can be left unattended and will not need turning. It makes an excellent soil conditioner.

also improve the growing conditions for certain plants, making them less susceptible to pests and diseases. Clubroot in brassicas is a good example, but it is important to bear in mind that by raising the pH it may reduce yield.

Various lime formulations are available. They all contain calcium, which is the active ingredient. Calcium carbonate is the most common and can be bought as crushed chalk or limestone. Quicklime can be used but it is extremely caustic and dangerous to handle. Slaked or hydrated lime is fast-acting but can scorch plant foliage.

It is preferable to apply lime to bare ground, at any time of year, but best in autumn or winter. Avoid doing so at the same time as animal manures, because the chemical reaction that occurs causes most of the useful nitrogen in the manure to turn into ammonia gas. Lime and animal manures should be applied separately, with a gap of at least several weeks. Gardeners are no longer encouraged to sprinkle lime onto their compost heaps for a similar reason: it depletes nitrogen levels.

Soil moisture and rainwater

All plants need a regular supply of water, which comes mainly from the soil, although some moisture can be absorbed through their leaves. So long as the garden soil is well prepared, moisture should be readily available, although during prolonged dry periods and periods of drought, supplementary watering may be needed. Plants growing in containers, where the roots are confined and not able to search for water, are far more prone to drying out and depend entirely on the gardener for all their water and nutrients.

Even when the soil is moist or wet, roots may not be able to access soil moisture and will start to wilt. This can happen in very heavy clay soils, where the clay particles hold the soil moisture very tightly; when the roots have been damaged by pests or diseases; or when the soil is waterlogged, which can lead to root rot and death.

Young plants and seedlings are far more prone to water stress and drying out, since they have a small, poorly developed root system that does not extend far down into the soil profile. Established plants, especially many trees and shrubs, have deep-growing roots that are able to search deeper into the soil for water and, as a result, are less prone to water stress. Shallow-rooted shrubs, however, such as heaths and heathers (*Calluna* and *Erica*), camellias, hydrangeas, rhododendrons and many types of conifer, are also prone to drying out and drought.

> **BOTANY IN ACTION**
>
> ## *Watering plants*
>
> When carrying out watering it is essential to water thoroughly. Light watering results in the water sitting in the top few centimetres of the soil, rather than moving deeper through the soil profile where the roots will benefit most. Regular applications of light watering often result in roots growing towards the surface, making them even more susceptible to periods of dry weather.
>
> Rainwater is usually better for plants than tap water, which may be too alkaline. The pH of rain varies, though it is typically acidic, primarily due to the presence of two strong acids – sulphuric and nitric – which may arise from natural or man-made sources.

Fully saturated soils

When soils are completely waterlogged or flooded, the water drives out the air in the soil and the plant roots are deprived of the oxygen they need to thrive. In winter, when plant roots are dormant, they can survive relatively long periods of waterlogging before suffering excessive damage, but in summer when the plant's water demand is high, even a few days can be fatal. Soils said to be at 'field capacity' contain their maximum amount of moisture before the air spaces are compromised.

Wilting point

The wilting point of a soil defines the minimal amount of moisture a soil can carry before the plants that it sustains begin to wilt. Conditions beyond this point could be described as drought, when the soil dries out during hot weather. During prolonged drought, plants may pass their permanent wilting point, where they can no longer recover their turgidity and will die. As soil moisture levels begin to decline to dangerously low levels, plants react by closing the stomata on their leaves to reduce water loss. They are rarely able to close completely, so they will always be losing water.

The first symptoms are wilting leaves, followed by leaf and bud death, and finally dieback of stems or the whole plant. The parts furthest from the roots are usually affected first and most severely.

An extreme response to drought seen in some trees is the sudden loss of entire limbs. This is frequently seen in the river red gum (*Eucalyptus camaldulensis*) of Australia, giving the tree its ominous nickname, widow maker. Limb dropping is a rare occurrence in the UK. However, during the very dry and hot summer of 2003, the large oak at the entrance to the RHS Garden Wisley suddenly dropped one of its limbs, falling onto the roof of the gift shop. No one was injured.

Drought tolerance and xerophytes

Many plants that naturally live in arid or very dry regions show numerous adaptations to protect themselves against dry weather and desiccation. They are called xerophytes.

Xerophytic plants may have less overall leaf surface area than other plants. They may have many fewer branches, as do barrel cacti, or small or reduced leaves. The extreme example of this is seen in cacti, where the leaves are reduced to spines.

Those xerophytes that do have leaves might be covered in a thick, waxy cuticle, or in tiny hairs that serve to absorb or trap water. The purpose of these hairs is to create a more humid atmosphere around the plant and reduce airflow, so reducing the rate of evaporation and transpiration. Yet others have scented leaves, the scent being produced by volatile oils. These oils, sitting on the leaf surface, help to prevent water loss, and it is an adaptation seen in many Mediterranean plants. Pale, silvery or white leaves reflect sunlight, and so reduce leaf temperature and evaporation.

Succulent plants store water in their stems or leaves, and sometimes in modified underground stems. Plants with bulbs, rhizomes and tubers, which are usually dormant during drought conditions, are not true xerophytes as they are drought evaders, lying dormant under the soil until the drought passes.

Many xerophytes have a specialised physiology, known as crassulacean acid metabolism. It is a modification of the standard photosynthetic process: the stomata only open at night (when temperatures are much cooler), and the plant is able to store carbon dioxide for photosynthesis during the day. The stomata may also be situated in pits, which makes them less exposed to the environment.

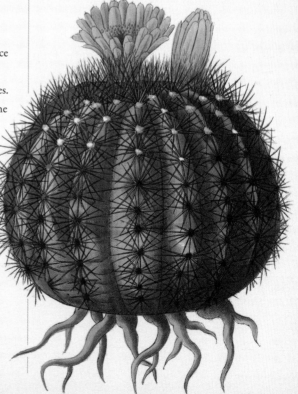

Echinocactus is a genus of barrel cacti, which do not branch; the leaves are reduced to spines to keep water loss to a minimum.

John Lindley FRS
1799–1865

John Lindley was an English botanist who was hugely influential in many aspects of British botany and horticulture and prominent in the early years of the Royal Horticultural Society (RHS).

Lindley was born near Norwich, where his father ran a nursery and commercial fruit business. Lindley loved plants; he would help his father wherever possible and spent his spare time collecting wildflowers. Despite his father's horticultural knowledge, the business was not profitable, the family was always in debt and Lindley was not able to attend university, as he had wanted. Instead, at the age of 16, he went to Belgium as an agent for a London seed merchant.

On his return to England, he met the botanist Sir William Jackson Hooker, who allowed him to use his botanical library. Hooker introduced him to Sir Joseph Banks, who employed him as Assistant Librarian in his herbarium. He also worked at Banks' house, concentrating on roses and foxgloves, and produced his first publications. These included *Monographia Rosarum*, which included descriptions of new species and some of his own botanical paintings, *Monographia Digitalium* and *Observations on Pomaceae*. These publications showed remarkable taxonomic judgment, detailed observations and precise use of language in both English and Latin, all despite his lack of university education. These, together with contributions to the *Botanical Register*, quickly won him international acclaim and a reputation as a brilliant botanist.

John Lindley was instrumental in the Royal Horticultural Society's inaugural years. The Society's Library in London is named after him.

During this time he met Joseph Sabine, a keen rosarian and Secretary of the Horticultural Society of London (which later became the RHS). Sabine commissioned him to produce botanical illustrations of roses, and in 1822 he was appointed Assistant Secretary to the Society at its new garden at Chiswick, where he supervised the collections of plants. He also organised a series of floral fetes, which became the first flower shows in England and the predecessors of the Society's famous flower shows.

Six years later he was elected Fellow of the Royal Society of London and appointed Professor of Botany in the newly founded University of London, where he wrote botanical textbooks for his students, as he was unhappy with the books that were available at that time. This was a position he held until 1860, when he became Emeritus Professor. He was unwilling to give up his employment with the Horticultural Society of London and held both positions simultaneously, becoming General Assistant Secretary in 1827 and Secretary in 1858. During this time, he carried a heavy load of responsibility for the Society and made important decisions during its financially troubled years.

Lindley always assumed responsibility for his father's heavy debts. Driven partly by financial necessity, and never being one to shirk a heavy workload, he took on ever more duties without relinquishing those he already had. In 1826, for example, he became *de facto* editor of the *Botanical Register* and in 1836, Superintendent of the Chelsea Physic Garden.

Lindley's expertise was also vital in many other important issues of the time. After the death of Sir Joseph Banks, the royal gardens at Kew went into decline and Lindley prepared a report on their management. Despite his recommendation that the gardens be turned over to the nation and used as the botanical headquarters for the United Kingdom, the government did not accept the findings. Instead, they proposed to abolish it and distribute the plants. Lindley raised the matter in Parliament, the government backed down and the gardens were saved, all of which led to the foundation of the Royal Botanic Gardens, Kew.

Lindley was also part of a scientific commission set up by the government to investigate potato blight disease and its effects that caused the Irish famine. The subsequent report helped bring about the repeal of the 1815 Corn Laws and helped alleviate the effects of the disease.

Lindley was acknowledged to be the top authority on the classification of orchids and his famous orchid collection was housed in the Kew herbarium. His *Theory and Practice of Horticulture* was considered to be one of the best books on the physiological principles of horticulture. He developed his own natural system of plant classification for his best-known book, *The Vegetable Kingdom* (1846). His large private botanical library became the foundation of the Lindley Library of the RHS. In 1841 he co-founded the periodical *The Gardeners' Chronicle*, which was published for nearly 150 years and he became its first editor.

Over his illustrious career, Lindley received numerous awards and commendations. He was elected a Fellow of the Linnean Society of London and a Fellow of the Royal Society of London, and was awarded an honorary degree of Doctor of Philosophy from the University of Munich.

He was highly regarded amongst the scientific community and the many species with the epithets *lindleyi* and *lindleyanus* are named in his honour.

The standard author abbreviation Lindl. is used to indicate him as the author when citing botanical names of plants.

Vanda sanderiana, waling-waling

Rosa foetida, Austrian briar, Austrian yellow rose

A hand-coloured engraving of *Rosa foetida* by John Curtis from John Lindley's *Monographia Rosarum*.

Nutrients and feeding

As discussed in chapter 3, plants need a wide range of mineral nutrients to complete all their living processes. As most of these are absorbed through the roots, the soil must contain the right amount of these nutrients in a form that is readily available.

In a natural setting, all of these nutrients must come from the environment, such as from the decomposition of leaf litter in woodlands. In gardens, however, the natural soil fertility may not be adequate to maintain healthy plant growth, particularly for some cultivars that have particularly high nutrient demands.

This means that supplementary feeding is often necessary. Lawns may also need feeding on a regular basis, as frequent mowing removes large amounts of nutrients that are not otherwise replaced.

Plants growing in containers are even more dependent on additional nutrients, as their roots do not reach out into the open soil. Good-quality potting composts usually contain adequate nutrients to feed the plants for five to six weeks, although some do contain long-term fertilisers that can feed for up to six months. After this time, it is up to the gardener.

Plants only need feeding when they are actively growing. Feeding plants when they are dormant can lead to a build-up of nutrients to toxic levels, which can damage or even kill the sensitive root hairs. Conversely, a lack of nutrients can lead to physiological disorders (see pp. 216–217).

Organic feeds

While some gardeners may regard most soil additives as organic feeds, the actual nutrient content of additives like home-made compost can be comparatively low. They are mainly added to the soil to improve its structure. True organic fertilisers contain relatively high amounts of nutrients, but because of their organic origin, they tend to release their nutrients over a longer period of time than factory-made synthetic fertilisers.

Organic fertilisers are often regarded as being better than man-made ones as they feed the soil, helping to maintain the soil microorganism populations, as well as feeding the plant. They also tend to use significantly less energy in their production. Feeds based on seaweed extracts, or those that contain seaweed, are useful plant 'tonics' as they contain a wide range of macro- and micronutrients, vitamins, plant hormones and antibiotics, although their nutrient content can be very varied.

> **BOTANY IN ACTION**
>
> ### Foliar feeds
>
> As well as absorbing nutrients through their roots, plants are also capable of absorbing them through their leaves, through the epidermis and stomata. As a result, some very soluble synthetic fertilisers can be used as a foliar feed, watering them over the plant as well as over the soil, where they will add to the total amount of nutrients absorbed. This method of feeding is often recommended where plants need to absorb nutrients quickly, often as an 'emergency' treatment. In some cases, such as tomatoes, it is also believed that foliar feeding during flower set produces a dramatic increase in fruit production.
>
>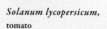
>
> *Solanum lycopersicum,* tomato

Life above ground

Outside the soil, plants are exposed to the atmosphere and the full force of nature: extremes of weather and temperature, rainfall, frost and strong or cold winds. Gardeners may seek to protect their plants in various ways (see chapter 7), but on the whole it pays to understand plants in the context of their natural environment. By doing so, gardeners are much more likely to succeed with their plantings.

Weather and climate

The term 'climate' refers to the weather patterns of an area over a long period, long enough to provide meaningful averages. 'Weather' describes the atmospheric conditions at a particular point of time. Or, as is often said, 'Climate is what you expect – weather is what you get!'

Microclimate

A microclimate is a localised zone where the climate differs from the surrounding area. In the great outdoors, this may be a sheltered microclimate on a south-facing cliffside (in the northern hemisphere) or an exposed wind tunnel formed by wind racing through a narrow valley. In a garden, microclimates include sheltered areas under trees, and sunny walls that retain heat in the day then release it at night, giving extra protection to plants growing near them.

Climate change

Climate change is a term that is often misused and misunderstood. It is often used in reference to global warming, the steady rise of the planet's average temperature – a real phenomenon that scientists have been recording over many decades. In fact, the planet has always been in a state of climate change; many thousands of years ago it was experiencing an ice age – one of several. Right now we are in an interglacial period, and the planet is warming up. All gardeners can do in response to climate change is to watch the weather and act accordingly!

Often regarded as a somewhat tender climber, *Campsis radicans* (trumpet vine, trumpet honeysuckle) is generally hardy down to -20°C (-4°F).

Temperature and hardiness

Both high and low extremes of temperature will adversely affect plant growth. The critical temperature above which plants are killed is called their thermal death point. Naturally, this varies from plant to plant; for instance, many cacti can survive very high temperatures, whereas shade-loving plants are killed at much lower temperatures. Temperatures above 50°C (122°F) will kill many temperate-region plants.

The lowest temperature that a plant can endure will determine its hardiness. Unsurprisingly, the range is enormous and it is usually divided into three very broad categories: tender, half-hardy and hardy. Tender plants are those killed by freezing temperatures, while half-hardy plants tolerate some degrees of cold. Hardy plants are well adapted to freezing temperatures, although some are tougher than others.

Unfortunately, the word 'hardy' has numerous connotations in gardening, and in some hot parts of the world it might refer to a plant's tolerance to drought or high temperatures. Hardiness is also relative – what is hardy in one country may not be hardy in another, and the term can also be applied so loosely that it is a poor guide. 'Hardy fuchsias' are a case in point: some can survive temperatures down to -10°C (14°F), while others will survive only brief periods below freezing.

Plant responses

Heat shock proteins

Plants react to heat by manufacturing specific heat shock proteins, which help the cells to function under periods of extreme temperature stress. Those plants that are acclimatised to high temperatures always have some heat shock proteins at the ready, so that they are able to respond quickly to further extremes.

Cold temperature stress responses

In temperate and arctic regions, it is temperatures at or below freezing point that affect most plants. In response, plants change their biochemistry; they may, for example, increase the concentration of sugars in their cells to make the cell solution more concentrated and less likely to freeze, so it stays liquid below the freezing point of water.

In very cold climates, such as within the Arctic Circle, plants actually dehydrate their cells and place the water removed from them between the cell walls, where it can freeze without causing damage to the cell contents.

These changes are known as 'cold hardening' and are triggered by the shortening days of autumn and colder temperatures. However, to fully acclimatise to freezing conditions, plants have to experience several days of cold weather before the freeze occurs, and this explains why even hardy plants can be damaged by a sudden autumn frost.

Plants also produce antifreeze proteins, which provide further protection against freezing temperatures by thickening the concentration of the cell solution. These proteins bind to ice crystals within the cell to prevent their development, which would otherwise rupture the cell.

Species of *Geranium* – such as *Geranium argenteum* (silvery crane's-bill) – often given the common name 'hardy geranium', will tolerate temperatures as low as -30°C (-22°F); other species are less hardy.

Frost and frost pockets

When the temperature of a surface reaches freezing point, water vapour in the air begins to deposit itself in the form of frost. When temperatures at night begin to start falling below 5°C (41°F) in the autumn, gardeners should expect the arrival of the first frosts and consider giving some protection to their frost-sensitive plants, such as dahlias and cannas.

Frost is particularly damaging to tender new growth and spring blossom. Depending on how low the temperate drops and how long the freezing period lasts, plants may start aborting buds, foliage, flowers and developing fruit to try and survive the conditions. Frost-tender plants, such as summer bedding, should not be planted out until night temperatures are safely above 5°C (41°F); in the UK this will not be until the end of May.

Plants growing in containers are particularly prone to cold and frost damage, because their roots are above ground level and not insulated by large masses of surrounding soil. Even otherwise hardy plants can be damaged or killed when grown in containers, when their entire rootballs can freeze. For this reason, many gardeners insulate their containers in winter, particularly if the container is small. This may also protect ceramic pots from cracking in the frost. However, insulation will not protect from prolonged or very low temperatures.

Frost pockets occur when very cold air travels downhill to collect in pockets in valleys and hollows or behind solid structures, such as fences or solid hedges. The reason this happens is that air becomes heavier as it gets colder.

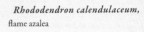

Rhododendron calendulaceum, flame azalea

Wind

Large plants like trees, or plants growing in exposed situations, are prone to wind damage during high winds and severe storms. Plants that naturally grow in exposed conditions may be dwarf or low-growing, or they may form mats, reducing their exposure to the wind. The typical image of a tree permanently bent to the wind's will is actually a response to exposure: its growth is concentrated on the leeward side where conditions are more sheltered.

Wind rips and tears at plants, sometimes causing large branches to fall, or even uprooting entire trees. Banana palms have 'rip zones' that allow the leaves to tear, reducing wind resistance and preventing the loss of the entire leaf or plant. Other plants may have small leaves to reduce wind resistance. Broad-leaved deciduous trees are especially prone to damage during freak summer or autumn storms; at this time their leaves are still attached to the branches and they really catch the wind. October 16th 1987 will be forever be remembered in British history as the night that a hurricane swept across southern England, destroying over 15 million mature trees, many of which had stood for centuries.

BOTANY IN ACTION

Tackling frost

Although frost can be damaging to plants on its own, repeated freezing and thawing, or rapid thawing can be particularly damaging. Camellias are particularly prone to rapid thawing, which is why they should never be planted where they will receive morning sun.

> ### BOTANY IN ACTION
>
> When the wind blows, the rapid movement of air removes large amounts of water from the leaves. Plants react by closing their stomata to reduce the amount of water lost. Grasses are able to roll their leaves to reduce the evaporation of water from their stomata, a response also seen during drought. Some plants may simply drop their leaves. If water cannot be replaced faster than it is being lost, wind-induced leaf scorch results, producing dry brown patches around the leaf margins.
>
>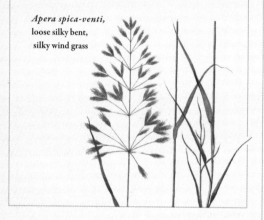
>
> *Apera spica-venti,* loose silky bent, silky wind grass

Rain, snow and hail

Although rainwater is essential to keep soils moist, torrential or heavy rain can adversely affect the tops of plants, breaking delicate stems, leaves and flowers. Continually moist conditions can also encourage the growth of fungi and other pathogens on plant surfaces. Some plants have long 'drip tips' on their leaves, which may be an adaptation to discourage water droplets from settling on their surfaces for any period of time. Long drip tips also create fine water droplets, and the finer a water droplet is, the less it will erode the soil around the roots of the plant.

Like rain, hail can break delicate plant parts; it is capable even of ripping through the leaves. It also causes bruising and abrasion and can defoliate plants or cause complete fruit loss. Small scars caused by hail on young leaves and fruit become very noticeable as the plants develop. For the fruit grower, hail is a curse.

Snow forms a source of moisture when it thaws, but before then it can cause damage to plants due to its heavy weight, which can put pressure on stems and break branches. A thick snow layer that sits on the ground for a prolonged length of time will starve plants of light and may cause them to die back or even die. A layer of snow on the ground can actually provide insulation for low-growing plants, protecting them from colder, freezing temperatures and winds. In alpine regions, herbaceous plants may even spring into growth before the snow layer has had time to melt; the new growth is protected by the blanket of snow, and by the time summer arrives, and the snow melts, the plants are already in flower.

Coastal conditions and salt

Continual exposure and wind-blown sand grains provide unique challenges to plants growing near the coast, and there is the added problem of salt in the air. When deposited on a plant, salt draws moisture out of plant tissues, causing desiccation and scorching. When the water droplets evaporate, the salt can penetrate stems, buds and leaves, causing direct damage and affecting the plant's cell structure and metabolic processes, resulting in bud death, stem dieback, stunted growth and, in extreme situations, the death of the whole plant. Road salt inadvertently applied to plants can have the same effect.

Plants of sand dunes and salt marshes are well adapted to the stresses caused by living in a high-salt environment, depending on how close to the shore they live. Plants that can withstand periods of inundation by sea water are known as halophytes. Since high salinity in the water makes it difficult for these plants to take up water, halophytes show a variety of physiological adaptations, such as rapid growth during times of immersion to reduce salt concentrations in individual cells, swelling of leaves to dilute the harmful effects of salt, or succulent leaves to retain water. Some mangrove trees deal with the continuous inflow of salt by carrying it away in their sap and depositing it in their older leaves, which are then shed.

Light

Gardeners should always take note of a plant's light requirements. Sun-loving plants grown in deep shade will quickly become weak and die, and shade-loving plants grown in full sun may scorch and dry out quickly and wilt. The factors that matter are the quantity of light, which is determined by day length, cloud cover and the amount of shade, and the quality of light, which may be of a certain spectrum (of importance to plants that grow on the forest floor, or to plants that grow in parts of the world where there is greater exposure to ultraviolet light).

Etiolation occurs when plants are grown in light levels that are too low for their natural requirements. Symptoms include overly long, weak stems, smaller and sparser leaves produced further apart, and a paler colour. Etiolation is typically seen in developing seedings grown on windowsills as they grow sideways and stretch towards the source of light. Many sun-loving plants also show similar symptoms when they are grown in too much shade. The tall flowering stems of *Veronicastrum virginicum*, for example, will grow out towards the sun if they are not given even light.

The effect of light on plants is also discussed in chapter 8.

Veronicastrum virginicum, **Culver's root**

Influencing the environment

There are many circumstances where gardeners can influence the environment to improve the growing conditions of their plants, or extend their season of growth to improve performance or increase yields. As well as keeping plants healthy by looking after the soil and growing plants in the right place, gardeners may use glasshouses, cloches, horticultural fleece and other forms of protected cultivation.

For example, horticultural fleece or bell cloches are used to start plants into growth earlier by keeping the air warm and still above their emerging growth, and glasshouses or polytunnels can be used to keep fruiting plants going for as long as possible before winter sets in (as with grapevines or chilli peppers).

Windbreaks and shelterbelts are used to shelter gardens from strong and cold winds. They can be highly effective in creating a mild environment conducive to plant growth, since they significantly reduce wind on their leeward side to a distance of 10 times their height. Windbreaks should be longer than the area needing protection, as wind can slip around the sides.

Thick mulches of organic matter are of great merit as they insulate the soil in winter, and in summer keep roots cool and reduce the amount of water lost through evaporation. Mulches also reduce weed growth, which would otherwise remove soil moisture and nutrients that would be available to the plants.

During winter and early spring, light levels can be too low to ensure strong, compact growth in young plants and seedlings, which can lead to etiolation. In these situations, gardeners can improve growth by providing supplementary lighting in the glasshouse or in the home, using lamps that provide light in the correct wavelengths: 400–450 nm and 650–700 nm.

Rosa,
rose

CHAPTER 7

Pruning

Pruning is the removal of material from a plant to improve its performance as well as its health, and to improve its overall look and size within a cultivated setting. Although it is a term that is usually reserved for woody plants, it can be loosely applied to any action that leads to the shedding of growth. This mainly includes pruning initiated by the gardener, such as deadheading, branch removal and clipping, but it can also include pruning induced by the plant itself: 'self-pruning' or abscission (see p. 161).

Pruning has been defined as a cross between an art and a science; a balance between knowing what to do and how to do it, and the overall aesthetic it will achieve. Of course, all plants can be left to develop without pruning, as happens in natural settings like woods and forests, but in a garden most plants will soon become untidy and overgrown.

Pruning works by affecting a plant's growth pattern. The hormone changes that it induces encourages dormant buds to shoot or new buds to form, and the change in the ratio between the amount of growth above ground in comparison to the growth below ground will induce its own particular growth response.

Why prune?

Pruning influences the way plants grow, and gardeners exploit these physiological reactions to improve their shape and their ability to produce flowers and fruit. Dead, diseased or damaged growth can also be removed by pruning, to improve the overall health of the plant.

Some plants need very little pruning; others need annual pruning to ensure they perform in a desired way. Overgrown plants that have never been pruned, or have been pruned badly, may need severe pruning to restore their shape and reinvigorate flowering and fruiting.

Many new gardeners worry unnecessarily that cutting off even the smallest twig will have dire consequences for the plant. In fact, most plants are very forgiving and respond well to pruning. Some even respond well to being cut back hard, and in such instances it may even prolong their lifespan. On the other hand, there are some plants that do not like being pruned much at all, such as rock roses (*Cistus*) and laburnums.

Pruning in a garden setting

Gardeners must never lose sight of the fact that a garden is a semi-natural environment. Left to nature it would quickly grow over, so the human element is vital. As well as weeding, pruning and cutting back are the keystones of garden maintenance. Primarily, it keeps plants within bounds and within the scale and proportion of the garden. An overgrown tree or shrub can soon become overdominant and ruin the balance of all other garden elements.

To this end, the early training of trees and shrubs is very important to create or maintain the required shape and habit. Weak, crossing, rubbing and overcrowded growth should also be removed on any plant with an easily apparent woody framework as it interferes with the visual appeal. This is not so vital for dense and bushy evergreen shrubs, like holly and bay trees for example, since their branches are not so apparent. Rubbing growth can damage the bark and lead to infection if not removed.

Dead, diseased, dying or damaged growth should also be removed, not only for aesthetic reasons, but also to prevent any infections from spreading or entering the plant. Large trees with diseased or dead branches may also pose a safety hazard and should be looked at and treated by a qualified professional.

Pruning can also improve flowering and fruiting, and it can be used to create visual effects of its own, such as topiary, stem and foliage effects (through coppicing or pollarding) and bonsai.

Grafted plants sometimes produce suckering growth from the rootstock, which must be removed as

Cistus salviifolius, sage-leaved rock rose

Ilex aquifolium 'Angustimarginata Aurea', golden variegated holly

and when it is seen, and many variegated plants have a tendency to produce growth with all-green leaves; this must also be removed promptly as it may come to dominate the plant.

Natural senescence and abscission

The process of deterioration that accompanies ageing, leading to the death of a plant organ, is known as senescence. The actual process by which these dead plant organs are shed is known as abscission. Both processes are seen every year when deciduous trees shed their leaves ready for winter, but they also apply to many other phenomena, such as flower drop and fruit fall.

Senescence often occurs seasonally. In annual plants, it happens every year as first the leaves die, followed by the stem and root systems; in biennials it occurs after two years. In perennial plants, the lifespan is indefinite, with the stem and root systems staying alive for many, sometimes hundreds of years, but with leaves, seeds, fruit and flowers being shed at different times.

Many evergreen species retain their leaves for only two to three years before they die and are abscised. In herbaceous perennials, leaf senescence progresses from old to young leaves, sometimes followed by the death of the entire above-ground portion of the plant before dormancy. Gardeners spend a lot of time during late summer and autumn tidying up these withered parts to keep the display refreshed; sometimes it may even stimulate fresh regrowth and further flowering.

There are a number of biological advantages of senescence and abscission. For many, the perpetuation of the species depends on abscising its fruits, which can then be scattered or transported to new locations. Old flowers and leaves may shade new leaves or become diseased if they are not removed, and the dropping of leaves helps to return nutrients to the soil – a sort of nutrient economy that helps forest trees to survive on infertile soils. Abscission is also seen on dehydrated plants that need to drop leaves to reduce transpiration.

BOTANY IN ACTION

Chlorophyll in leaves

The senescence of autumn leaves provides an extra ornamental effect in many trees, shrubs and some perennials, with the characteristic change in colour from green to reds, yellows and oranges. As the daylight hours reduce and temperatures drop in autumn, the chlorophyll levels in the leaves decrease. This reveals the otherwise hidden or masked yellow xanthophylls and orange-yellow carotenoid pigments present in the leaves. At the same time, leaves may synthesise anthocyanins, which impart the red and purple colours.

Look for autumn colours in *Acer platanoides* 'Aureovariegatum', Norway maple.

Physiological responses to pruning

In order to prune correctly it helps to understand how plants grow – and how they respond to pruning. New shoot growth is usually made from the uppermost bud on a stem, called the apical or terminal bud, and plants show different bud arrangements along their stems: either opposite, alternate or whorled (see pp. 67–68). Timing is important if pruning is not to interfere with flowering and fruiting. Such knowledge in respect to a particular plant not only helps a gardener know how and where to make the cuts, but also when to prune.

Apical dominance

The apical or terminal bud imposes a controlling influence, known as apical dominance, over the growth of buds and stems below it and on lateral growth or sideshoots. When the apical bud is removed, this dominance is lost and buds lower down start to shoot. Gardeners exploit this response to create bushier plants, and in its extreme, regular clipping can be used to create fine topiary.

In some plants, if the apical bud is removed, a single sideshoot may grow away strongly and reinstate dominance; in others, two or more growing points share the dominance and produce dual or multiple shoots. In trees, multiple leading shoots can lead to problems later on, so the weakest should be removed to retain the strongest and / or straightest.

Apical dominance is controlled by the terminal bud, which produce the plant hormone auxin (see p. 98). For a shoot tip to be active, it must be able to export auxin into the main stem, but if substantial amounts of auxin already exist in the main stem, an export channel cannot be established and the shoot remains inactive. All the shoot tips compete with each other, so that tips both above and below can influence each other's growth. This allows the strongest branches to grow the most vigorously, wherever they may be. The main shoot dominates mostly because it was there first, rather than because of its position at the tip of the plant.

Pulling down a vertical shoot or branch and training it horizontally can also break apical dominance. Sideshoots that are produced along this shoot are much more likely to flower and fruit. This technique is particularly useful when training climbers, wall shrubs and some styles of trained fruit tree.

Branching patterns

Plants can be categorised by the bud arrangement on their stems. This determines their leaf and branch arrangement: either alternate, opposite or whorled. Opposite buds, and hence leaves and stems, are in pairs at the same level on opposite sides of the stem. Alternate buds alternate at different levels on opposite sides of the stem. Whorled buds produce a circle of three or more leaves or shoots at each node. (See p. 68 for leaf and branch arrangement.)

The terminal bud and shoot suppress the growth of lateral buds

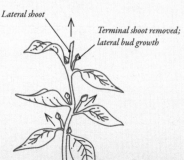

Lateral buds are stimulated when the dominant bud is cut off

To prove the role of auxin, growth of lateral buds is suppressed if terminal shoot replaced with an agar block that has been soaked in auxin.

The shoots that emerge from buds tend to grow in the direction the bud was pointing. Pruning above an alternate bud will induce a new shoot that grows in the direction the bud points, whereas pruning above a pair of opposite buds produces two shoots, one either side of the remaining stem.

Plants with opposite shoots are harder to prune, because it is hard to squeeze the tips of the secateurs into the V to cut immediately above the bud so as not to leave a stub to die back. Also it is harder to direct the growth of opposite-shooted plants towards the desired direction – one stem will grow in the right direction (out from the centre, perhaps) but is accompanied by one going in the opposite direction. One solution is to rub off the unwanted bud or to go back several weeks later and cut off the unwanted shoot that results.

Timing of pruning

Correct timing is essential to ensure the best flowering and fruiting displays. Weather and climate, however, also play their role, so a gardener needs to be sensitive to the needs of each plant. If some evergreens and slightly tender plants are pruned too early in spring, or too late in autumn, for example, the resulting cuts or new growth may be susceptible to frost or cold wind damage.

Some plants will bleed sap profusely if pruned at the wrong time of year, which will weaken or even kill them. This applies to birch trees in summer, grapevines in spring and walnuts in winter. Other plants may be more susceptible to disease if pruning is carried out at the wrong time; for example, all stone fruit of the cherry genus (*Prunus*) are at risk from bacterial canker and silver leaf disease if pruned when dormant in winter; wait until summer before pruning.

A plant's response to pruning depends not only on how much you cut off a stem but also on when you do it. Pruning a dormant stem removes buds that would have grown into shoots or flowers, and since the food reserves within the plant are then reapportioned among fewer buds, the shoots from the remaining buds grow with increased vigour. As the growing season progresses, a plant's response to pruning changes, with its response becoming less vigorous after the midpoint of summer.

For woody plants that are grown for their flowers, the timing of pruning is determined by their season of interest. A gardener will need to know if the flowers form on the new, current year's wood, or whether they arise from the old wood that was formed during the previous summer. As a general rule, plants that flower on new growth will not do so until after midsummer, as the early part of the season is spent forming the new growth. Plants that flower on old wood are able to do so much earlier in the season, and generally do so anytime from midwinter until midsummer.

Prunus avium,
bird cherry

Pruning trees

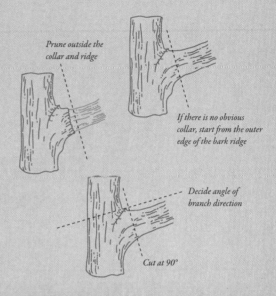

Tree collars

In trees, the branch structure and how they are attached to each other falls into three categories: collared unions (which contain a branch collar), collarless unions, and co-dominant unions (where two branches of roughly the same diameter emerge from the same point, with neither showing dominance over the other). A branch collar contains tissues that provide a protective chemical zone, which compartmentalises the cut area and inhibits the movement of infection.

Each type of attachment is best tackled in a particular way, so that the branch has the best chance of sealing over and compartmentalising decay.

Branch unions

Collared unions need to be cut just outside the raised bark ridge and swollen collar area where the branch joins the trunk; do not cut the collar off or cut through it. Collarless unions have no swollen branch collar and should be cut along an angled line starting from the outer edge of the branch-bark ridge, running away from the trunk. With co-dominant branches, the top of the cut is made just beyond the wrinkled bark of the branch-bark ridge, ending on the outer side of the limb, where it meets the main branch.

In practice, the branch-bark ridge and collar are sometimes difficult to find, making it difficult for amateurs to make an accurate assessment. At other times, the positions of branches can make access for the saw either difficult or impossible. Heavy limbs may also tear the bark as they fall away, causing further damage that is difficult to remedy without further amputation. If a large tree is to be pruned, always hire a professional with the correct tools, experience and qualifications – it is much better this than face an accident life-threatening to either yourself or the tree.

Girdling

As mentioned on p. 65, girdling, or bark ringing, can be used to control a plant's growth. It has a similar effect to root pruning and is a very useful process for unproductive apple and pear trees, but not stone fruit. The aim is to remove an incomplete 6–13 mm ($^1/_4$ –$^1/_2$ in) wide ring of bark from the main trunk at a suitable height above ground level, but well below the lowest branches. The cut must penetrate right through the bark and cambium layer.

If an incomplete ring is made with about one-third of the bark left intact, the tree will still be able to draw water and minerals up from the roots, and the flow of sugars and other nutrients will still be able to flow down to the roots, but in a much-reduced supply. As a result, vigour should be greatly curbed, although if the job is done incorrectly it may result in the tree's death.

Bark ringing should only be used as a last resort and on very vigorous trees, and it is carried out in mid- to late spring. Notching or nicking a small crescent of bark just above a bud is another technique similar to girdling, but with a much more localised effect. The purpose is to stimulate the bud just below the cut into growth.

Pruning

Constant tidying up, snipping off shoots and lightly tipping back plants may lead to unbalanced growth that is top-heavy or lopsided, and if it is done with no regard to correct timing it may also remove developing flower buds and so result in no floral or fruit display. Some gardeners may hack back hard at plants (hard pruning) while others may be so timid that they are reluctant to do anything more than a light trim.

To be effective, gardeners need to know which of these two approaches is most appropriate, and when they need to take an intermediate approach. Be aware that many plants do not respond well to hard pruning because they may not have dormant or adventitious buds and so will not re-shoot from old wood or old growth.

Contrary to what might be expected, pruning hard is not always the answer to reducing the size of an overgrown plant, since such plants often respond to hard pruning by sending out more vigorous growth, although this depends on the size, health and vigour of the root system. The renovation and reduction in size of large, overgrown trees is usually done, therefore, over several years of selective (but not hard) pruning. Strong stems are usually tipped back lightly by one-half or one-third; weaker stems can be pruned out completely.

Pruning cuts

Clean, sharp tools must be used at all times when making pruning cuts to encourage fast healing and to reduce the risk of infection. It is also important to make the right cut in the right place, at the right time.

The pruning cut should be clean with no ragged edges or stem damage and be as close to the bud as possible, but not so near that it damages it. The aim is to ensure that the stem ends in either a strong bud or a healthy sideshoot.

Wisteria sinensis,
Chinese wisteria

Wisterias need pruning in summer and winter to build up short spurs on which the flowers form and to curtail excessive growth.

Any growth that remains beyond the remaining bud or sideshoot will simply die back, since there are no buds or any other organs above it that need to be sustained. The resulting 'snag' or 'stub' that remains may be a source of disease infection, which may travel back through the rest of the stem.

Cut healing

After pruning, the plant reacts by producing an accumulation of proteins in the exposed cells, temporarily protecting the underlying tissues from infection. Xylem and phloem vessels may also produce antifungal compounds. The cut surface then begins form a callus.

A callus is a mass of unorganised parenchyma cells that start to cover the cut surface. It forms from the vascular bundles and cork cells and then grows inwards until the whole area, or at least the outer ring of the cut, is covered over. Individual cells may also undergo changes to form 'walls' around the wound to prevent any potential infection from spreading.

Prunus domestica (plum). Using a wound paint after pruning *Prunus* species may help as an extra precaution against silver leaf disease.

It is often suggested that large pruning cuts made on branches should be covered with wound paint. As a rule, this practice is no longer recommended as these paints have been found to hinder the natural healing processes of callus formation. They can also seal in moisture and make a better environment for pathogens. The only exception is with trees of the genus *Prunus*, such as plums and cherries, where wound paint is thought to be useful as an extra precaution against silver leaf disease.

Root pruning

By this process, it is possible to reduce and restrain excessive vegetative top growth, although lifting and pruning a tree has far more practical difficulties than branch pruning. Root pruning reduces the plant's vigour and, as a result, promotes the formation of flowering growth instead of leafy growth. This makes it very useful for improving unproductive flowering shrubs and fruit trees. It is carried out from autumn to late winter, when the tree is fully dormant.

Young plants – up to five years old – can simply be dug up and the roots pruned before being replanted. Older trees up to ten years old need more extensive preparation by digging a trench around the tree, 30–45 cm (12–18 in) deep and wide, and about 1.2–1.5 m (4–5 ft) from the trunk. The major roots are then severed and the trench refilled as soon as possible. Older and mature trees should not be root-pruned, except as a very last resort, as they have much less resilience than younger trees.

Pinching out and disbudding

New soft growth can be 'pinched out' using just a thumb and forefinger. This reduces extension growth from that point and encourages the formation of sideshoots and enhances bushiness. It is often carried out on seedlings and young plants to prevent them becoming tall and spindly, but pinching out can be carried out on any plant with soft growth.

Disbudding is the removal of surplus buds. This is normally performed on fruit trees to prevent a plant from forming too many flower buds and subsequently too much fruit. It is used to control the amount and quality of fruit. The process can also be used on plants grown purely for their flowers; it presupposes that a plant can only produce a certain total quantity of flowers, so if bigger flowers are desired, their number must be reduced. This applies not only to the size, but also to the quality, since the available nutrients and water are divided between fewer flowers.

Remove sideshoots from cordon tomatoes to encourage upright growth and crops of larger, better fruit

Deadheading

The process of removing flowers when they are fading, dying or dead is known as deadheading. It is done to keep plants looking attractive (as dead flowerheads are often untidy in appearance) and, where plants have the potential to do so, it encourages them to make a further display of flowers.

Once the flowers have been pollinated, seed and fruit development begins, and this often sends signals to the rest of the plant so that further flower development is retarded. Regular deadheading prevents seed and fruit being produced and prevents energy being wasted on their production. This extra energy is redirected into producing stronger growth and sometimes further flowers.

Plants that are commonly deadheaded include bedding plants, bulbs and many herbaceous perennials. While bulbs do not produce further displays, deadheading will prevent them from wasting energy on seed production, ensuring the energy is used instead for the following year's flowers. Remove the developing seed capsule but leave the green flower stalk in place until it dies, as this produces food through photosynthesis.

It is not often necessary to deadhead flowering shrubs, since many are pruned each year after flowering anyway (see later, p. 172), nor is it practical. Some shrubs, however, particularly those with large flowers, will benefit from deadheading, and these include camellias, rhododendrons, lilacs (*Syringa*), roses and tree peonies.

Some gardeners talk of 'liveheading' their plants. It is practised on a number of late-summer herbaceous perennials to encourage later and bushier displays. The emerging flowerheads are cut back once in early to midsummer (no later) and allowed to regrow. Plants that respond to liveheading are asters, phloxes and heleniums.

BOTANY IN ACTION

Post-pruning feeding

After pruning, it is a good idea to feed the plants with a fertiliser. As plants store a lot of food reserves in their stems, cutting them off reduces the overall amount of stored reserves and it is important to encourage good regrowth after pruning.

Liquid feeds will give a quick boost, but are short-lived; granular feeds last longer; controlled-release fertilisers feed for several months. Most flowering trees, shrubs and climbers respond well to feeding with a high-potash granular feed.

As a plant's feeder roots are mainly distributed at the edge of the leaf canopy, the feed should be spread in a circle in this area. Some gardeners erroneously apply the feed at the base of the stems where there are few feeder roots to make use of it.

After feeding, water the soil to activate the fertiliser and add a thick mulch of organic matter, such as well-rotted manure, garden compost or composted bark, to keep the soil moist. See p. 152 for more about plant fertilisers.

Faded flower and developing seed head

***Helenium*,** sneezewood

Marianne North
1830–1890

Marianne North was an English botanical artist and naturalist. She travelled widely to satisfy her passion for painting flora from all around the world.

Initially, she had planned to be a singer and trained as a vocalist, but sadly her voice failed. As a result, she changed career aspirations, painting flowers and pursuing her ambition of painting the flora of other countries.

From the age of 25 she travelled extensively with her father, but after his death, which had a profound effect on her, Marianne decided to travel further by herself. Her globetrotting was inspired by these earlier travels and the plant collections she saw at the Botanic Gardens, Kew.

She began her travels at the age of 41, first going to Canada, the USA, Jamaica and then on to Brazil.

Then, in 1875, after a short spell in Tenerife, she began a two-year journey around the world, painting the flora and landscapes of California, Japan, Borneo, Java, Ceylon and India and capturing them in oil paint.

Fortunately for Marianne she was well connected, thanks to her father's political career, and made use of contacts who supported her in her travels, including the President of the USA and the poet Henry Wadsworth Longfellow. In the UK, Marianne had many supporters including Charles Darwin and Sir Joseph Hooker, then Director of Kew.

Despite all these famous connections, Marianne preferred to travel unaccompanied and visited many areas unknown to most Europeans. She was at her happiest when discovering plants in the wild, surrounded by their natural habitat, and painting them. Some of the plants Marianne painted were new to science and some were named after her.

On her return to Britain, she had several exhibitions of her botanical paintings in London. She then offered her complete collection of paintings to the Royal Botanic Gardens, Kew, and even proposed to build a gallery to house them. The Marianne North Gallery first opened in 1882 and is still open today. It is the only permanent solo exhibition by a female artist in Britain. Visitors to Kew Gardens can enjoy this Victorian treasure house in all its glory and view the remarkable botanical art collection of this pioneering painter.

Even then, her travelling days were not over. In 1880, at the suggestion of Charles Darwin, she went to Australia and New Zealand and painted there for a year. In 1883, following the opening of the Marianne North Gallery, she continued to travel and paint, visiting South Africa, the Seychelles and Chile.

Marianne North was a Victorian artist who travelled many parts of the world recording with her paintbrush the flora she found there.

> 'I HAD LONG DREAMED OF GOING TO SOME TROPICAL COUNTRY
> TO PAINT ITS PECULIAR VEGETATION ON THE SPOT IN
> NATURAL ABUNDANT LUXURIANCE...'

The remarkable scientific accuracy of her paintings gives her work a permanent botanical and historical value. Her paintings of *Banksia attenuata*, *B. grandis* and *B. robur* are very highly regarded. A number of plant species were named in Marianne North's honour, including *Crinum northianum* (a synonym of *Crinum asiaticum*), *Kniphofia northiae* and *Nepenthes northiana*, as well as the genus *Northia* (family *Sapotaceae*), endemic to the Seychelles.

Wild pineapple in flower and fruit, painted by Marianne North in Borneo in 1876 during her two-year journey around the world painting flora and landscapes.

North studied briefly under the Victorian flower painter Valentine Bartholomew. This study of *Nepenthes northiana* demonstrates her vigorous style.

THE MARIANNE NORTH GALLERY

The Marianne North Gallery at Kew contains 832 brilliantly coloured paintings of flora, fauna and scenes of local people from around the world, which Marianne North carefully recorded in just 13 years of travel. It celebrates Marianne North's contribution as an artist and explorer, as well as one of Kew's first benefactors.

Unfortunately, some years ago the Gallery building and the paintings inside started to suffer. Unlike modern exhibition spaces, the building had no proper environmental controls, so heat, damp and mould were damaging both the building and the paintings. The roof was no longer sound and the walls were not weathertight. Thankfully, in 2008 Kew obtained a substantial grant from the National Lottery, which enabled it to mount a major restoration project of both the gallery and the paintings. This included new touch-screen interactive installations.

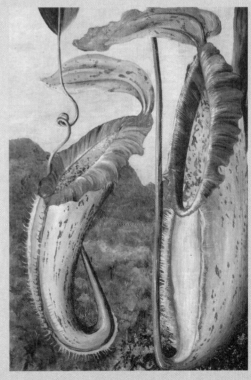

Pruning for size and shape

Very few woody plants can be left alone in a garden setting without some sort of intervention from the gardener. Most importantly, it keeps them at the right scale. Many trees and shrubs will become too dominant and untidy if allowed to grow unchecked, and rampant climbers become a heavy, tangled mess.

Even if a natural look is required in a garden, the illusion of minimal effort still needs careful attention to pruning. Light pruning after flowering will keep any wayward growth or overly vigorous plants in check, and shrubs that might block out views can be discreetly contained. Some larger shrubs like *Buddleja davidii* might need hard pruning every year in late winter so that they remain contained and can spring back to life each summer.

Plants that require repeated pruning throughout the year are probably too large for the space that they are growing in. In such instances, it may be better to remove the plant and replace with something more suitable. Common examples include eucalyptus trees, the spring-flowering *Clematis montana* and Leyland cypress trees (× *Cuprocyparis leylandii*), which are often planted as hedges but soon become too tall.

Removing dead, diseased, dying or damaged growth

Dead growth should always be completely removed – it serves no use to the plant and is a potential source of infection. Anything that is damaged, diseased or starting to die back should also be removed. All diseased material should be promptly removed from the site of infection and destroyed.

Pests that can be wholly or partially controlled by pruning include aphids (especially woolly aphids), red spider mites, stem-boring caterpillars and scale insects. Diseases that can be controlled by pruning include brown rot, canker, coral spot, fire blight, mildew, rust and silver leaf. Gardeners should not rely on pruning as a method of control for any pest or disease, however, since their presence often indicates an underlying cause; the best advice is always to find the source of the problem and treat that (see chapter 9).

If the damaged or diseased growth has already healed naturally by itself, then it is usually better to leave this in place rather than trying to cut it out to new, healthy growth. Branches that are partially broken, such as by strong winds, grazing animals or lightning strikes, rarely mend if tied back in position and it is usually better to cut them out or shorten them to a suitable replacement stem.

Damaged or torn bark is most unlikely to reunite with the rest of the tree.

Buddleja davidii,
butterfly bush

Formative pruning

Many trees and shrubs naturally develop a well-shaped branch structure. Others will benefit from formative pruning in the first year or two, especially if they are of a species that only needs minimal pruning when mature. Very young trees, particularly whips (those with no or few side branches), will definitely need formative pruning to build up a good branch structure, as will young fruit trees and bushes.

Formative pruning of shrubs that are pruned hard annually is not so important, but it still helps to produce a balanced shape and symmetry with well-spaced stems by removing branches that are crowded or crossing. After pruning, the plant should have evenly spaced branches and a balanced, open habit. This forms the permanent woody framework from which the bulk of the plant develops. As the plant continues to grow, aim to maintain this open habit with the new growth.

Pruning evergreen shrubs

Evergreen shrubs generally need the minimum of pruning apart from deadheading (where manageable), removal of any dead branches or shoots, and clipping or trimming to shape. If possible, broad-leaved evergreens are best trimmed with secateurs, since shearing can unattractively shear their larger leaves, which will develop brown edges. Bear in mind that pruning too early in the year will produce young growth that may be susceptible to spring frost damage, and that pruning too late in the year, such as late summer or autumn, produces soft growth that will not have time to harden before winter sets in. Gardeners in the northern hemisphere use the timing of the RHS Chelsea Flower Show in late May as a guide. Pruning at this time is not too early and not too late; it is known as the 'Chelsea chop'.

Rosmarinus officinalis, rosemary

Rosemary should never be hard-pruned as it will not re-shoot from old wood. Instead, trim plants annually.

Some evergreen shrubs positively hate being pruned hard. As they do not produce many adventitious buds, they do not readily re-shoot from old wood. These include brooms (*Cytisus* and *Genista*), heaths and heathers (*Calluna* and *Erica*), lavender and rosemary. If these shrubs are left without pruning for several years and then pruned back hard into old, brown wood, they will simply die or produce very poor, weak and spindly regrowth that looks very unattractive and never performs well. Instead, these plants should be lightly pruned or trimmed annually to keep them strong and bushy, to the required size and flowering well.

Pruning for display

Pruning is one of the main ways gardeners can influence how well woody plants flower. Its timing mainly depends on when the plant flowers, and whether flower buds are produced on the new, current year's growth or on wood that ripened in the previous summer.

Penstemon gracilis, beardtongue

Flowers

Plants that flower any time from late autumn to late spring (sometimes early summer) will have to have produced their flower buds during the previous summer and so they are borne on 'old' wood. Plants that flower during summer and early autumn, when the plant is actively growing, produce their flowers on the current year's growth. The general rule for pruning flowering shrubs is to prune them as soon as the flowers fade. This will allow a whole season for the next year's buds to form. Shrubs that flower late in the summer or autumn, such as hydrangeas, are often left until spring before they are pruned, to give a degree of frost protection. It is usually a good idea to wait until growth buds start to shoot before pruning.

Late-flowering shrubs that are on the borderline of hardiness, such as fuchsias, penstemons and cape fuchsias (*Phygelius*), can be damaged by frost or cold weather if pruned too early and they are best left until mid-spring or later if cold weather persists.

Fruit

The primary aims of pruning for fruit are to promote flower / fruit bud production, to increase sunlight penetration into the crown to aid ripening, to remove less-productive growth, and to shape the crown into an efficient, stable form. If left unpruned, the overall quantity of fruit produced might be greater, but the size and quality of individual fruit much less.

Many fruit trees have two types of buds: the flower buds that will go on to form the fruit and the vegetative buds that will produce leafy growth and new stems. On many tree fruit, the flower buds are fatter than the vegetative buds, and they form on short shoots or spurs along the length of the branches. These are called spur-bearing trees. Where the flower buds are clustered at the tips of the branches, the tree is known as a tip-bearer.

Malus domestica, apple

Correct pruning of apple trees will build up wood that produces a regular and bumper crop of good-sized fruit.

The growing conditions during the season of bud initiation and the subsequent winter weather will affect the number of buds that flower. Flower buds only usually appear on two- or three-year-old stems, so it can take several years of training before a young tree will produce a significant crop.

Biennial bearing

Some fruit trees, especially apples and pears, experience biennial bearing – they only produce a decent crop every other year. There are numerous factors that affect this, including the health and vigour of the tree and environmental factors, and some cultivars are more prone than others.

Trees may become stuck in a pattern of biennial bearing, producing huge crops in an 'on' year and very little in an 'off' year. In an 'on' year, the tree puts such a huge strain on its water, nutrient and energy supplies that it takes the whole of the next season to recover. It is possible to overcome this biennial pattern by careful pruning in the winter following the 'off' year – prune as normal but leave on the tree as many one-year-old shoots as possible. These will not produce fruit the next year, but will instead produce flower buds for the year after that: the next 'off' year. Another method is to thin out the blossom, removing nine out of every ten trusses within a week of flowering.

BOTANY IN ACTION

Water shoots

As a result of excess pruning, several buds can break at the same point and time, producing a thicket of thin stems called water shoots. In time, these may produce useful stem regrowth and go on to flower and fruit normally, but they are usually so numerous that they crowd one another out. Water shoots should either be thinned out or removed completely if not needed to produce replacement growth.

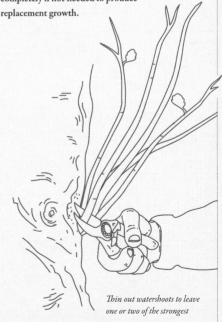

Thin out watershoots to leave one or two of the strongest

Cornus alba,
red-barked dogwood

Pruning for stems

A few deciduous shrubs, notably the dogwoods, *Cornus alba*, *C. sanguinea*, *C. sericea*, ornamental *Rubus* (brambles) and some willows (*Salix*), are grown for their colourful stems, useful for the winter garden. As these stems age, the colour intensity fades, so they are hard-pruned in spring to encourage new stems that colour up strongly by the autumn.

This pruning is usually carried out annually, although it can be reduced to every two or three years if necessary. If the height of the shrub is integral to the summer border, simply remove just one-third to one-half of the oldest stems each year, with this rotation of pruning carried out over a two- to three-year period.

This technique of cutting back hard to ground level is also known as coppicing or stooling. Dogwoods and willows can also be pollarded, where the stems are encouraged to grow from the top of a trunk, the height of which can be varied to suit the gardener's requirements. Note that not all trees and shrubs will respond to this treatment.

Coppicing

This technique of cutting back all the stems of a shrub or tree to just above ground level is particularly useful to promote colourful juvenile stems and ornamental foliage as well as to rejuvenate plants that tolerate hard pruning. Overgrown hazels, hornbeams and yews (one of the few conifer species that will tolerate hard pruning), can be cut close to the ground in late winter. This results in the production of lots of new stems that can be reduced in number and thinned out to make an open, more airy bush once more.

Hazels, in particular, are coppiced every few years to produce long, straight stems that can be used in the vegetable garden as bean sticks. Coppicing is a very severe, but useful, way of keeping some trees that naturally grow tall within bounds. For example, coppicing will keep *Paulownia tomentosa* to the size of a large shrub, where it can easily grow 3 m (10 ft) in a single year, with the curious side-effect of producing extra-large, very ornamental leaves. Some *Eucalyptus* can be treated similarly, especially as tall trees are very susceptible to wind rock and being blown down in strong winds.

1. Large or overgrown shrub suitable for coppicing.

2. Hard-prune stems in winter or early spring.

3. Strong new shoots will be quickly produced.

4. Excess stems may need to be thinned out.

Pollarding

All the stems of a tree or shrub can be cut back to the top of a single trunk – a pollard. It is a useful technique for creating a taller structural feature at the back of a border and for restricting the height of certain trees like alders (*Alnus*), ashes (*Fraxinus*), tulip trees (*Liriodendron tulipifera*), mulberries (*Morus*), plane trees (*Platanus*), oaks (*Quercus*), limes (*Tilia*) and elms (*Ulmus*). The best time for pollarding is in late winter or early spring and it is best started on young trees as young wood responds rapidly to wounding, reducing the risk of decay.

BOTANY IN ACTION

How to create a pollard

Initially the tree is allowed to grow to the desired height and then the side branches are removed. Stems are encouraged to sprout from the top of the trunk. Each year, these new stems are cut back to the top of the trunk, and over time the trunk, or pollard, will become thicker and more established.

Once pollarded it is important to continue the cycle of cutting as the weight and angle of the new branches can lead to weakness, particularly where many stems are crowded together.

Each time the tree is pollarded, stems should be cut above the previous pollarding cuts, to avoid exposing older wood, which may be at an increased risk of decay.

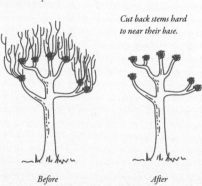

Cut back stems hard to near their base.

Before   After

Pruning for foliage

Many deciduous shrubs and trees are grown mainly for their large or colourful foliage, and good examples include the Indian bean tree (*Catalpa bignonioides* and its yellow-leaved cultivar 'Aurea'), purple-leaved hazel (*Corylus maxima* 'Purpurea'), the smoke bush (*Cotinus coggygria* 'Royal Purple' and other cultivars), the foxglove tree (*Paulownia tomentosa*), stag's horn sumach (*Rhus typhina*) and various elders (*Sambucus nigra*). These shrubs will produce larger leaves if hard-pruned annually in spring.

Unfortunately, this hard pruning will more or less prevent the plants flowering, which for most of these shrubs is of no consequence. If you do want them to flower, then either prune every three of four years or allow them to develop without hard pruning.

Corylus maxima,
filbert, large filbert

Narcissus tazetta,
bunch-flowered daffodil,
Chinese sacred lily

Chapter 8

Botany and the Senses

In the words of David Attenborough, the famous broadcaster and natural historian, 'Plants can see. They can count and communicate with one another. They are able to react to the slightest touch and to estimate time with extraordinary precision'. This may seem fanciful, but the more that botanists learn about plants, the more they discover just how intimately plants are able to sense their environment.

Experienced gardeners will already have some inkling of this, but probably not to the extent to which it actually happens. Most people just think of plants as more or less inanimate objects, and this view can be forgiven because the time scale of plants is quite different to our own.

Most of us already know, however, that emerging shoots seek out the light, and that germinating seeds are able to sense gravity. Everybody knows that some flowers turn their heads to the sun, and some close at night. Others can even catch prey or fend off their enemies. The more we discover, the more examples we find. The sensory capability of plants is a phenomenon that botanists are only just beginning to understand.

Seeing light

It has been known for many years, since the very early experiments in the mid-17th century, that plants sense or 'see' light and react to its presence by starting to photosynthesise (see pp. 89–90). Nowadays botanists are able to describe a number of ways that plants respond to light: photomorphogenesis, which is the way a plant develops itself structurally in response to light; phototropism, which is the method by which plant tissues grow away or towards light; and photoperiodism, the synchronisation of plant growth with time.

Visible light is just one part of the whole electromagnetic spectrum, which includes x-rays, gamma rays, radio waves and microwaves. Sound, incidentally, is not part of the spectrum. Light wavelength is measured in nanometers (nm, billionths of a metre), and it is the one part of the electromagnetic spectrum that the human eye can see. Each colour making up the visible spectrum has a different wavelength – red has the longest wavelength (620–750 nm) and violet has the shortest (380–450 nm); green has a wavelength of 495–570 nm. When all the colour wavelengths are seen together, they make white light. Either side of the visible spectrum are infrared light (750–1000 nm) and ultraviolet light (300–400 nm). Plants are able to detect these wavelengths too.

Plant organs contain photosensitive compounds (photoreceptors) that react to the presence of light, and even specific wavelengths of light. The main photoreceptors are phytochrome (absorbs red and blue light), cryptochrome (absorbs blue and ultraviolet light), UVR8 (absorbs ultraviolet light) and protochlorophyllide (red and blue light). From this we can see that the red and blue wavelengths of light, at opposite ends of the spectrum, are most useful to plants. Interestingly, the only colours that our eyes see are those that are reflected back to us from an object. We perceive plants as being green because these are the wavelengths of light that the plant does not absorb.

Phototropism

Stems and leaves frequently orientate themselves in respect to the main source of light. Roots seldom exhibit phototropism, but they do tend to bend away from light if they come into contact with it. Growth towards the light is known as positive phototropism, growth away from it is known as negative phototropism.

The experiments of Charles Darwin and his son Francis, in 1880, demonstrated that the phototrophic stimulus is detected at the plant's growing tips, but that the bending of the tip is caused by cells lower down. To show this, they used the coleoptiles (the protective sheath around the emerging shoot tip of monot seedlings) of germinating oat seedlings. Those that had their tips covered were unable to respond to the light, but those that had only their tips exposed were still able to grow towards it. It took the work of the Danish scientist Boysen-Jensen in 1913 to show that some sort of chemical signal was being sent from the tip of the plant to cells further down. This led to Frits Went's later discovery of the plant hormone auxin in 1926.

An increase in light intensity leads to a corresponding increase in phototropism – but if the light becomes too intense, there will be negative phototropism, i.e. the plant will start to retreat from the light. High light levels, particularly in the ultraviolet range, may also stimulate the manufacture of anthocyanins – a type of natural sunblock. Plant scientists have demonstrated that phototropism is triggered by both the red and blue parts of the light spectrum, leading them to believe that more than one type of photoreceptor is responsible for phototropism.

Photoperiodism

The synchronisation of plants with seasonal and daily time is known as photoperiodism. It causes many plant responses, such as stem elongation, flowering, leaf growth, dormancy, stomatal opening and leaf fall. It is also widely seen in animals. In fact, much of what we see in the natural world is happening because plants and animals are able to detect the varying lengths of day or night.

Seasonal photoperiodism plays a greater role the further away plants are from the equator, where there is very little seasonal variation in day length. In cool-temperate regions, for example, seasonal differences are huge; entire forests of trees drop their leaves as the summer fades and herbaceous perennials die back to their ground-dwelling buds.

The rate at which day length changes varies during the year. Near the summer and winter solstices there is little variation in the rate of change, but during the spring and autumn equinoxes, day length changes more rapidly. Often it is the length of darkness, rather than that of daylight, that plants are responding to.

In plants, the study of the effects of photoperiodism has mostly been on flowering times, as this usually correlates with season. Under controlled conditions (i.e. with the vagaries of weather taken out of the equation) a given species of plant will flower at approximately the same date each year.

Plants are either long-day plants (LDP) or short-day plants (SDP). LDPs flower only when the day length exceeds their critical photoperiod, and these plants typically flower during late spring or early summer as days are getting longer. SDPs only flower when the day length falls below their critical photoperiod, and these are typically late summer or autumn-flowering plants such as asters and chrysanthemums. Natural nighttime light, such as moonlight, or that from street lighting is not sufficient to interrupt flowering.

Day-neutral plants, those that are unaffected by day length, are sometimes encountered. Many common weeds are day-neutral, but so are broad beans and tomatoes. They may start to flower after reaching a certain overall developmental stage or age, or in response to different environmental stimuli, such as a period of low temperature.

BOTANY IN ACTION

Some species of plant will only flower if the photoperiod is just right (such as beetroots, turnips, lettuces and peas), while others are less fussy and will flower even if the photoperiod is not quite reached (such as strawberries, rye grass, oats, clovers and carnations). A few days of the required or approximate photoperiod are usually required to initiate flowering, although some plants, such as Japanese morning glory (*Ipomea nil*), need only one short day to initiate their flowering cycle. Evidence suggests that a hormone named florigen is responsible for floral induction, but its existence is a matter of dispute.

Convolvulus tricolor, morning glory

Photomorphogenesis

Plant response to light that is neither directional nor periodical is known as photomorphogenesis. It is how light causes a plant to develop. An example is seen during germination, when the emerging shoot first encounters light. It will send a signal down to the root, causing the root to start branching. Plant hormones are an important part of photomorphogenesis, as they are the signals that one part of a plant will send out to initiate a response elsewhere. Examples might be tuber formation in potatoes, stem elongation in low light, or leaf formation.

Colour signals

Colour is used by plants, often to trigger the senses of animals. No gardener can deny being attracted to plants with plenty of large and colourful flowers. In the wild, colourful flowers are used to attract pollinators, acting like shining beacons.

Pollinators respond differently to the different wavelengths of light, and flowers are coloured specifically to attract their pollinators. Many insects, particularly bees, respond to long wavelengths of light in the blue, violet and ultraviolet range, whereas plants predominantly pollinated by birds will have flowers coloured red and orange. Butterflies prefer colours such as yellows, oranges, pinks and reds.

Many flowers are patterned with streaks or lines, called nectar guides. These serve as landing strips for insects, directing them towards their nectar or pollen rewards. Some nectar guides are visible under normal light conditions, but many only show up under ultraviolet light. Fluorescence is also seen occasionally, perceptible in low-light conditions.

The colour of most flowers fades after successful pollination and usually it is the intensity of their colour that diminishes. This acts as an indicator to any passing pollinator that the flower has aged, that there is no pollen or nectar reward, and that they should move on to another flower. In some plants the flower actually changes colour after pollination, as in certain members of the *Boraginaceae* family, such as forget-me-nots (*Myosotis*) and lungworts (*Pulmonaria*), which change from pink to blue. Colour changes are also seen in fruit to indicate their ripeness.

'Touching' and 'feeling'

Plants are not only sensitive to touch, but they are also sensitive to other external forces such as gravity and air pressure. The directional response to touch is known as thigmotropism, and the response to gravity is known as geotropism.

Thigmotropism

The tendrils of some climbing plants, such as species of grape (*Vitis*), are strongly thigmotropic. Their tendrils feel the solid object on which they are growing, by detecting the contact via sensory epidermal cells called tactile blebs or papillae, which results in the coiling response. Any stems that twine around a support, or any clinging roots or twining petioles, are also doing so by thigmotropism.

The plant hormone auxin once again plays an important role. The cells that have received the physical stimulus produce auxin, which is transported to the growth tissue on the opposite side of the shoot that has been touched. This tissue then grows faster and

Pulmonaria,
lungwort

Passiflora alata,
winged-stem passionflower

BOTANY IN ACTION

The effects of wind and stroking

Plants growing in windy situations produce thicker and tougher stems to resist wind damage. This has a practical consideration for gardeners when it comes to staking trees. A short stake no longer than one-third of the height of the tree, rather than a longer stake reaching up to just below the lowest branches, produces a tree that is free to flex its trunk in the wind and encourages thickening near the base of the trunk. This results in a tree that is able to fend for itself when the stake is removed. Brushing over young seedlings with your fingers or a sheet of paper produces stronger stems and therefore more resilient seedlings and young plants.

elongates to bend around the object. In some cases, the cells on the contact side compress, which further enhances the curving response.

Roots exhibit a negative thigmotropic response, growing away from objects they feel. This allows them to grow through the soil, taking the path of minimum resistance, and to avoid stones and other large obstacles.

The leaves of the sensitive plant (*Mimosa pudica*) are famous for being able to close up and droop when touched. This is not a thigmotropic response, however, but rather a form of thigmonastism – a similar phenomenon but resulting from a different mechanical response.

Thigmonastism is an immediate response based on very fast changes in cell turgor (how much water the cells contain). It is not caused by cell growth. Other thigmonastic responses include the closing of the traps of the Venus flytrap (*Dionaea muscipula*) when insects land on them, and the twining of the stems or roots of the strangler fig (*Ficus costaricensis*).

Geotropism

As with their work on phototropism, Charles Darwin and his son were also the first to demonstrate this phenomenon. Sometimes referred to as gravitropism, geotropism is the growth reaction of plants in response to gravity. Roots exhibit positive geotropism, growing downwards in the direction of the Earth's gravitational pull, and stems exhibit negative geotropism, growing upwards in the opposite direction.

In the garden, geotropism is seen in germinating seedlings, when the emerging root begins its downward exploration of the soil. However, it is also seen in some trees and shrubs with a weeping or exaggerated downward growth habit. Examples include weeping pear (*Pyrus salicifolia* 'Pendula') and Kilmarnock willow (*Salix caprea* 'Kilmarnock'). Plants such as *Cotoneaster horizontalis* produce stems that seek to grow along the ground, rather than up or down – a sort of neutral geotropism.

PIERRE-JOSEPH REDOUTÉ
1759–1840

Belgian painter and botanist Pierre-Joseph Redouté is probably the most recognised of all botanical artists. He is especially renowned for his watercolours of lilies and roses painted at the Château de Malmaison in France. His paintings of roses are said to be the most often reproduced flower paintings of any botanical artist. He was often referred to as the 'Raphael of flowers'.

Redouté lacked formal education and he followed in the footsteps of both his father and grandfather, leaving home in his early teens to become a painter, carrying out portraits, religious commissions and even interior decoration.

When he was 23, he joined his elder brother, Antoine-Ferdinand, an interior decorator and scenery designer, in Paris. There he met botanist René Desfontaines, who steered him towards botanical illustration. At this time, Paris was considered the cultural and scientific centre of Europe, and interest in botanical illustration was at its height and his career flourished.

In 1786, Redouté began work at the Muséum National D'histoire Naturelle, where he prepared illustrations for scientific publications, cataloguing the collections of flora and fauna, and even took part in botanical expeditions. The following year, he went to study plants at the Royal Botanic Gardens, Kew. Here he also learned the art of stipple engraving and color printing, which provided him with the technical expertise needed to produce his beautiful botanical illustrations. He later introduced stipple engraving to France. He also developed a method of colour application that involved using small chamois leathers or cotton mops to apply a succession of colours to a copper engraving.

In subsequent years, he was employed by the French Academy of Sciences. It was during this time that he met the Dutch botanical artist Gerard van Spaendonck, who became Redouté's teacher, and French botanist Charles Louis L'Héritier de Brutelle, his patron, who taught him to dissect flowers and portray their diagnostic characteristics. Van Spaendonck taught him how to paint using pure watercolour. He developed a skill at translating his observations of nature onto paper that was very much in demand.

L'Héritier also introduced him to members of the court at Versailles, and Marie Antoinette became his patron. As her official court artist, he painted her gardens. Other patrons included the kings of France, from Louis XVI to Louis-Philippe. Redouté continued to paint right through the French Revolution.

The works of Pierre-Joseph Redouté are some of the most famous botanical illustrations.

Rosa cultivars, roses

This illustration is in *Choix des plus belles fleurs*, Redouté's compilation he considered to be among the best of his career.

In 1798, Empress Joséphine de Beauharnais, the first wife of Napoleon Bonaparte, became his patron and eventually he became her official artist, painting the plants in her Château de Malmaison garden as part of her project to develop and document the gardens. He even accompanied Bonaparte on his Egyptian expedition. It was under the patronage of the Empress that Redouté's career flourished and he produced his most sumptuous work, portraying plants from places as far apart as America, Japan and South Africa.

After the death of the Empress, Redouté became the master of design at the Musée d'Histoire Naturelle, and ended up giving drawing classes to female royals and painting for aesthetic value, including his book *Choix des plus belles fleurs*.

Redouté collaborated with the greatest botanists of the time and provided illustrations for nearly 50 publications. His principal works include *Geraniologia*, *Les Liliacées* (8 volumes), *Les Roses* (3 volumes), *Choix des plus belles fleurs et de quelques branches des plus beaux fruits*, *Catalogue de 486 Liliacées et de 168 roses peintes par P.-J. Redouté* and *Alphabet Flore*.

During the French Revolution, he documented numerous gardens that became national property. Over his career, Redouté contributed more than 2,100 published plates depicting more than 1,800 different species.

A large number of Redouté's original watercolors on vellum are in the collection of the Musée National de la Malmaison in France and other museums, although many reside in private collections.

Sprekelia formosissima, Aztec lily, St James lily

This illustration was also used in *Choix des plus belles Fleurs*, although it is labelled as *Amaryllis*.

Sensing scent

As seen in chapter 3, plants will react to certain molecules in the air, the best-known of these being ethylene, which causes a response in ripening fruit. Ethylene is also known to play a role in senescence of plant organs, such as flower drop and leaf fall, and it is known to be produced from all parts of higher plants. The chemical has also been found to be released in response to environmental stresses, such as flooding and wounding.

Studies show that when insect pests attack plants, the plants release into the air a variety of volatile chemicals, including pheromones, which are picked up by neighbouring plants. One of these chemicals, methyl jasmonate, induces nearby plants to start producing organic chemicals, particularly tannins, that will help them fight off and resist the impending attack (see pp. 210–211). The smell of cut grass, lovely as it is, is actually caused by the release of a variety of volatile chemicals. These may have a plant defence function, and they may be a signal that can be picked up by other plants.

Dodder (*Cuscuta*) is a parasitic plant whose seedlings use chemosensory methods to detect and grow towards their hosts. A series of experiments by scientists at the Penn State University in the USA were used to prove that one of the species (*C. pentagona*) grows towards volatile chemicals produced by tomato plants. They began by showing that if dodder is drawn to the tomato plant, a favourite host, the seedling will lean and grow towards it – even in the dark. They then demonstrated that dodder grown close to a fake plant with no tomato odour will just grow upright. Faced with a choice between a tomato plant and a wheat plant, dodder was shown to always grow towards the tomato. Finally, to prove that dodder smells the tomato, the two plants were grown in two separate boxes connected only by a tube through which the chemicals produced by the tomato could reach the dodder. The dodder grew towards the tube.

Furthermore, when grown next to a wheat plant only, dodder will grow towards it as it needs a food source, but if given the option, it will always grow towards tomatoes. It was demonstrated that wheat produces one attractant that the dodder can detect, but tomatoes produce a cocktail of three chemicals that are more attractive to dodder.

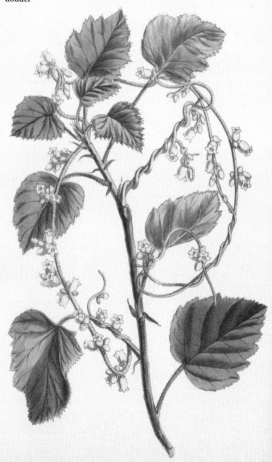

Cuscuta reflexa, dodder

SCENT AS AN ATTRACTANT

The smell of some flowers remains fondly in our memories for many years and these memories are instantly recalled when we smell the plant again. Some plant smells are so strong that they can be overpowering, and some plants produce a fragrance that can only be described as disgusting! Small wonder that it is plants that give us the world's most beautiful perfumes.

Most of the plant scents we know and recognise are those produced by flowers. As with extravagant colours, fragrance is mostly used in flowers to attract pollinators, and in fruit to demonstrate ripeness. Volatile compounds are produced, which readily evaporate. Different species contain different oils that combine together to produce the plant's distinctive scent. One of the most abundant and volatile compounds is methylbenzoate.

The scent of many winter-flowering plants can be very strong indeed. Garden favourites such as winter honeysuckle (*Lonicera fragrantissima*), *Daphne bholua* and Christmas box (*Sarcococca confusa*) are well known; their scent is so strong because of the scarcity of pollinators in winter – they must advertise themselves as widely as possible to attract pollinators from far away.

Lonicera fragrantissima, winter honeysuckle

Those plants that produce scented foliage, mainly do so to deter grazing herbivores (see pp. 210–211). Scented foliage is also found on many plants from hot and dry climates, examples include those used as culinary herbs such as rosemary, basil, bay and thyme. As well as a means of plant defence, these 'essential oils' also play an important role in protecting the foliage against drought and subsequent desiccation. A layer of oil or oil vapour on or around the surface of a leaf serves to reduce water loss.

Dictamnus albus (burning bush) is a good example of a plant that produces scented foliage. It is native to open habitats in southern Europe and northern Africa, and in hot weather it produces so many volatile oils that they can be ignited with a match, giving the plant its common name.

Feeling vibrations

Although it may sound unlikely, recent research shows that plants can respond to certain sounds and vibrations.

The role of sound in plants has yet to be fully explored, but the theories behind such scientific studies propose that for plants to know what is happening in their surrounding environment, it would be advantageous for them to use sound, since acoustic signals are found everywhere in their natural environment.

Sound perception

The flowers of numerous plant families demonstrate one common example of soundwave perception: they only release their pollen when their anthers are vibrated at the correct frequency – a frequency achieved by visiting bees that have co-evolved to vibrate their flight muscles accordingly.

The roots of young maize plants (*Zea mays*) have been shown to emit audible and frequent clicking noises. These roots also react to specific sounds by exhibiting frequency-selective sensitivity, which causes them to bend towards a sound source. Similarly, young maize roots suspended in water grow towards the source of a continuous sound emitted in the region of 220 Hz, which is within the frequency range that the roots emit. Other research shows that sounds can produce changes in germination and growth rates.

Sound production

While this sounds fanciful, it has been known for some time that plants produce and emit soundwaves at the lower end of the audio range, within 10–240 Hz, as well as ultrasonic acoustic emissions, ranging from 20 to 300k Hz. The mechanisms by which plants produce sounds are poorly understood.

Plant cells vibrate as a result of the active movement of organelles within them. Vibrations from individual cells are then propagated as soundwaves to neighbouring cells. If the receiving cells are receptive to this frequency, they will start vibrating too. If the sound extends outside the plant, it may behave as a signal.

> ### BOTANY IN ACTION
>
> #### Talking to plants
>
> There are some gardeners who believe that talking to their plants does improve their growth. There is no proof that the soundwaves are responsible for this. Some argue that carbon dioxide and water vapour in exhaled breath might improve plant growth, but the temporarily raised levels are unlikely to persist long enough near the plant to have any significant effect.

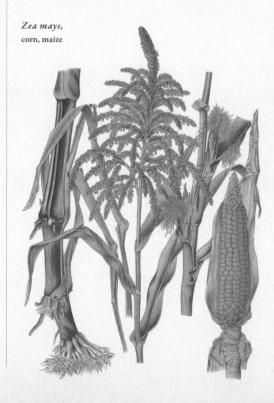

Zea mays, corn, maize

Human senses and plants

It's not just about the plants, though – gardeners have feelings too. Some plants prosper because they have attributes that gardeners find desirable. A garden might contain plants that satisfy all our senses: sight, hearing, touch, taste and smell.

Sight and sound

We grow plants to produce beautiful displays, usually to satisfy our sense of vision. If a plant looks beautiful to us, we grow it. The plant satisfies our demand by producing ornamental features, mainly flowers, fruits and foliage. A gardener, however, can create more depth by adding other sensory dimensions, such as sound and smell – a particularly attractive option for those who are visually impaired.

Gardeners often overlook the sounds generated by plants, but the rustling noise of foliage swaying in a gentle breeze is very relaxing. The sounds produced by grasses and bamboos in particular are one of the major assets of this group of plants.

Sound, however, can be a nuisance, where gardens are close to major roads, railway lines, industrial estates and other sources of disturbing sounds. In such instances, plants can be used to create a living sound barrier. Such barriers are often more effective than walls or fences as they absorb, deflect and diffract the sound waves and vibrations. They will also help prevent wind damage in the garden and mitigate traffic pollution.

A single layer of trees has very little effect as a noise barrier. Inter-reflection between plants means that a double row is far more effective and the overall effect can be a noise reduction of up to 8 ddB (decibels). In most situations, you can reduce the noise level by 1.5 dB for each 1 m (3 ft) of plant height. Although this reduction may be small, the perception of noise is dramatically decreased by vegetation, as it is far more acceptable if the source of the noise cannot be seen.

Although evergreens are considered the most effective screen, possessing foliage all year round, many deciduous trees and shrubs are effective during the summer when they are in full leaf, and this tends to be when we are outside in the garden. A 'fedge', sometimes called a willow wall, is a cross between a fence and a hedge. A fedge consists of living willow, woven into two parallel walls up to 4 m (13 ft) high. The space between the two walls is filled with soil. A fedge can reduce noise levels by up to 30 dB, and can be more effective than concrete walls, fencing or soil banks.

Touch

To many gardeners, touch can be an important plant attribute. The feel of a plant can vary from soft to very hard, depending on the leaf structure. Leaves may be leathery, hairy, smooth or shiny, and this range of textures can add significant variety and interest to the garden.

Taste and smell

We would probably not grow many of our garden herbs if they did not smell or taste nice. Take the flavour out of basil (*Ocimum basilicum*), for example, and there would be nothing left to make it such a remarkable plant. Few plants in the vegetable garden have much to offer in the way of ornament, but many are loved for their fresh flavour, which is sometimes hard to match in the shops.

Likewise, take scent away from a garden and a huge component of its beauty would be lost. Although it is hard to quantify scent, the smell of a garden can often overwhelm us on a warm spring day: the smell of freshly cut grass, the lingering scent of jasmine early on a summer's morning, or the cool afternoon air of a sunny autumn day. The smell is often crucial when we buy them. When we are looking to buy a rose, for example, it is scent that is often the overriding consideration – and, sadly, not all roses have a strong scent.

Malus floribunda,
Japanese crab apple

Chapter 9
Pests, Diseases and Disorders

To all herbivorous organisms, plants are food. They form a massive energy resource that is readily exploited by all plant-eating animals, particularly insects. The impact of herbivory on plants, therefore, cannot be underestimated.

The enormous destructive potential of herbivores, however, has failed to prevent plants from dominating the Earth, and we rarely see the total defoliation of plants by pests – in the wild, at least. The defensive strategies of plants, which have evolved in response to the attack strategies of herbivores, play a major role in their survival.

Likewise, disease epidemics are equally rare. Far more common are isolated incidents where individual plants that are already weak or stressed succumb to a bacterial or fungal infection. In modern times, however, introduced pests and diseases can wreak havoc with native and garden flora, as plants are brought from one part of the world to another, bringing with them infected material of non-native origin. Currently in Britain, ash trees are seriously under threat by a fungal disease known as ash dieback (*Chalara fraxinea*), and there is much concern about a fungal-like disease *Phythophthora ramorum*, which is a threat to Japanese larch trees (*Larix kaempferi*) and trees in the beech family (*Fagaceae*) as well as many ornamental shrubs.

Insect pests

The impact of insects on agriculture and horticulture has been documented for thousands of years. The Bible mentions them in the book of Exodus, and throughout the 1800s pioneering farmers in North America were plagued by terrifying swarms of locusts that would consume all plant material in their path. It is only in the last 100 years, however, that scientists have begun to understand the herbivorous interactions between plants and animals.

In terms of evolutionary history, herbivorous insects have evolved and adapted in tandem with host plants and natural enemies. It is not surprising, therefore, that various species of insects have developed different host-plant associations coupled with differing life cycles and feeding mechanisms – all necessary for the exploitation of their hosts.

Any plant species will, through evolutionary time, have 'gathered' its own specific insect fauna, many of which have become closely associated with particular plants. In the garden, the codling moth (*Cydia pomonella*) of apples and pears is a good example, as is the red lily beetle (*Lilioceris lilii*) – in both instances their type of host forms part of their Latin names. Other pests, such as some aphids, are more generalist feeders. Each of the different species that exploit a particular host plant do so in a number of unique ways (see opposite page).

Plants as food for insects

In terms of its composition – mostly water and lots of indigestible compounds such as cellulose and lignin – the majority of plant tissue is a fairly low grade of food. To acquire the nutrition they need, therefore, insect herbivores must consume large quantities of plant for each unit of insect growth. In other words, they need to eat a lot, or, in the language of science, adopt a 'high-consumption strategy'.

Some plant tissues such as seeds, pollen and nectar offer a better food source than others. Rapidly growing young tissues, full of active meristems, are a good source of food, but few insects use them as this is a bad feeding strategy – it may kill the host plant. Some insects, such as aphids, tap into the phloem and feed on the sap; this does not usually cause direct damage to the plant's productive machinery (unless it results in infection by a virus or severe leaf curling, for example), it simply removes some of the photosynthetic product.

Plants are known to produce barriers to insect feeding, making them unpalatable, poisonous or physically difficult to access (see p. 209). Mutualistic associations with herbivorous insects are also seen in

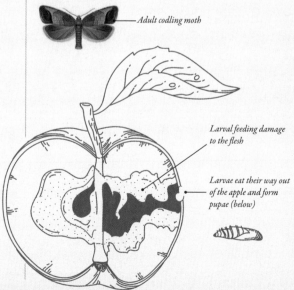

Adult codling moth

Larval feeding damage to the flesh

Larvae eat their way out of the apple and form pupae (below)

Female codling moths lay their eggs on or near developing fruit. After hatching, the caterpillars bore into the fruit and feed on it, tunnelling and causing a lot of damage.

some plants, such as in the relationship between the ant *Pseudomyrmex ferruginea* and the bull's horn acacia (*Acacia cornigera*) in Central America. The ants are fed special protein secretions by the tree, and in return the ants viciously protect their host from other herbivores, and they also destroy any other plants that come into contact with their host. This extraordinary association was first described by the evolutionary ecologist Daniel Janzen in 1979, and it gives the tree an enormous advantage over its competitors.

How insects feed

Plant-eating insects have evolved a variety of different feeding mechanisms for exploiting plant tissue, and the varied structure of their mouthparts tells this story. They can be classified broadly into two types: those for biting and chewing, and those for piercing the plant and sucking sap.

The mouthparts of chewing insects consists of mandibles, maxillae and a lower lip. In sap-sucking insects, these basic mouthparts have undergone severe modification with the mandibles drawn out into a long and thin, needle-like structure called a stylet.

As mentioned earlier, a single species of plant may be exploited by a number of different insects that all feed in their own particular way. The rosebay willowherb (*Chamaenerion angustifolium*), for example, can play host to a surprising number of insect herbivores. Its leaf chewers include larvae of the elephant hawk moth (*Deilephila elpenor*) and those of the moth genus *Mompha*, as well as flea beetles of the genus *Altica*. Its sap suckers include aphids (which feed on the leaf and stem phloem), the psyllid bug *Craspedolepta nebulosa* (which feeds on root phloem), the capsid bug *Lygocoris pabulinus* (which feeds on leaves) and the common froghopper *Philaenus spumarius* (which feeds on the stem xylem). Chewing insects have mouthparts that are modified to eat or chew through plant parts, mainly leaves, but also flowers, stems, buds and roots. The pair of mandibles, one on each side of the head, is used to either cut, tear, crush or chew their food. The commonest chewing insect pests are the larvae of many species of beetles, sawflies, butterflies and moths (caterpillars), together with some adult beetles.

Leaf miners are the larvae of insects that are tiny enough to live inside leaves, between the upper and lower leaf surfaces. As they eat through the leaf, internal tissue, they leave a mine showing their progress, hence their name. Most leaf-mining insects are species of moths, sawflies or flies, although the larvae of some beetles and wasps are also leaf miners. Leaf-mining species are often specific to one group or genus of plants, such as the holly leaf miner (*Phytomyza ilicis*) or the chrysanthemum leaf miner (*P. sygenesiae*).

Sap-sucking insects have a needle-like mouthpart, called the stylet, which pierces the epidermis of the

The leaves of rosebay willowherb are the usual food of the larvae (caterpillars) of the elephant hawk moth, but they also eat other plants such as bedstraw.

plant and draws out the sap from either the phloem or xylem. With the exceptions of leafhoppers, which have wider stylets, no visible holes are made in the plant tissue, although many sap suckers produce toxic saliva that causes leaf discolouration or distortion, such as leaf curling. This leaf curling provides the insects with some degree of protection against natural predators. Some sap suckers, especially aphids, leafhoppers, thrips and whitefly, can transmit virus diseases in their saliva.

Most sap suckers feed on above-ground parts of the plant, most commonly the leaves, although there are some species of aphids, mealybugs and some planthoppers that feed on the roots. Common sap-sucking insect pests include aphids, leafhoppers, mealybugs, scale insects, thrips, psyllids and whiteflies.

Honeydew is a sugar-rich liquid, secreted by sap-sucking insects as a waste product. It is collected by some species of wasps and bees. Some ants 'farm' aphids in order to collect their honeydew, even moving them to parts of the plant where sap production is at its strongest. Honeydew also attracts sooty mould, a powdery black fungus that can cover plant parts. Sooty mould does little harm to the plant; it is essentially a cosmetic problem, although it can reduce the amount of sunlight reaching the leaves. Cars parked under trees that are covered with honeydew-excreting insects may soon become dirty with sooty mould.

Aphids

These are by far the biggest group of insect pests in the garden. They are sometimes called greenfly and blackfly, depending on their colour, although species vary from yellow to pink to white or mottled. They range in size from about 2–6 mm (1/16–1/4 in).

Infestations of aphids are easily seen with the naked eye, and they tend to colonise shoot tips, the underside of leaves and even flower buds, often in huge numbers. Some species, such as woolly aphids (*Eriosoma lanigerum*) and beech woolly aphids (*Phyllaphis fagi*), cover themselves with a fluffy

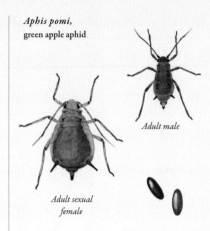

Aphis pomi, green apple aphid

Adult male

Adult sexual female

Eggs, which turn from green to black after they are laid

white waxy secretion, which protects them against desiccation and from some predators.

Aphids can reproduce at an alarming rate – a few aphids in spring can soon become huge colonies of many thousands, although many of these end up as food for birds and predatory insects. For most of the year, aphid colonies consist of wingless female nymphs that feed on their host plant. After a series of moults, the females become adults and give birth to live daughter aphids by parthenogenesis, without the need for a male participant. The daughters grow quickly and in as little as 8–10 days may begin to reproduce themselves. One female hatched in spring can produce up to 40–50 generations of females through a typical summer.

Some female nymphs mature into winged adults and are able to fly to infest new plants, although they are not strong flyers and are carried from plant to plant mostly by chance. In late summer and early autumn, winged male aphids begin to appear and these mate with the female to produce eggs. Most aphid colonies die out over winter and it is these eggs that ensure their survival from year to year. In spring, when female nymphs begin to hatch out of the eggs, the cycle begins again.

Secondary hosts

Some insect herbivores need two plant species to complete their life cycle. This can include cultivated garden plants as well as weed species, so the control of weeds can sometimes help to reduce infestations. Some aphids spend the whole year on one type of plant, although they may only be active for part of the year.

The black bean aphid (*Aphis fabae*), for example, lays eggs that overwinter on shrubs including spindle (*Euonymus europaeus*), mock orange (*Philadelphus*) and *Viburnum*. Its secondary hosts, on which it spends the summer, include a number of crops such as beans, carrots, potatoes and tomatoes, as well as more than 200 different species of cultivated and wild plants.

The plum leaf-curling aphid (*Brachycaudus helichrysi*) lays its overwintering eggs in the bark and on the buds of *Prunus* species (such as plums, cherries, peaches, damsons and gages). The aphids feed on the leaves, producing the characteristic leaf curling. In late spring or early summer, winged adults fly to various herbaceous plants, particularly those in the family *Asteraceae*, where they spend the summer. In autumn, winged adults fly back to the trees to lay the overwintering eggs.

Pest outbreaks

Some insects may cause great damage. In ecological terms, such outbreaks are often the result of a simple predator–prey imbalance, for example when the frequency of a plant species has become overabundant in a certain locality. This would tally with the experience of any farmer, or vegetable or fruit gardener: plants of any particular type grown in a relatively high density are usually more susceptible to pest attack than those grown in isolation. Conversely, it is usually easier for the gardener or farmer to manage such outbreaks if the susceptible crop is grown in a high-density area.

There are three types of pest outbreak: pest gradients, pest cycles and pest eruptions. Pest gradients are a sudden increase in pest density, often triggered by an abundance of food material in a small area. Once this has become exhausted, the pest population either has to migrate or it undergoes a massive population crash.

Pest cycles are periodical outbreaks that follow the season of the plant host, its food source. Crop rotation seeks to break the cycle of soil-borne pests, as well as those of diseases, by moving crops of a certain type to a different location each year. The large pine weevil (*Hylobius abietis*) shows cyclical outbreaks in conifer plantations, whenever there is clear felling. Before the practice of clear felling became commonplace, this pest was never known to exhibit cyclical behaviour. Outbreaks are much less common in forests where there is natural regeneration instead of clear felling.

Pest eruptions are perhaps the most damaging. Outbreaks suddenly occur and then spread out into the surrounding area after being dormant for a long time. After the outbreak, insect populations may remain at a higher level for some time before reverting to their normal level.

Viburnum opulus, guelder rose, is one of the winter hosts for the black bean aphid as well as other aphids. They feed on the newly opened spring buds and leaves.

Other common pests

While gardens are vital homes to wildlife, there are some species that do so much harm that they are difficult to fully appreciate.

Slugs and snails

Long-time foes of the gardener, these soft-bodied molluscs shouldn't really be classified as pests as they play an important role in recycling decaying organic matter. Unfortunately, some species can cause great damage to seedlings and the new growth of herbaceous plants in spring.

Most slugs and snails feed at night, and often produce tell-tale slime trails which alert you to their presence. Damage is most severe during warm or mild damp periods. Because of the protection provided by their shells, snails can move more freely over drier terrain than slugs. Unlike snails, slugs remain active throughout the year; snails lie dormant during autumn and winter.

Most species live in or on the soil surface, but keeled slugs (*Milax* species) live and feed mostly in the root zone and can cause severe damage to underground plant parts, such as roots and tubers. Reproduction occurs mainly in autumn and spring, when clusters of spherical, creamy-white eggs can be found under logs, stones, pots or in the soil.

Spider mites

These are not true insects; spider mites are members of the mite subclass, and they have eight legs rather than the six legs of insects. Spider mites are less than 1 mm (1/32 in) in size and vary in colour. Many species spin a silk webbing, and it is this webbing that gives them the 'spider' part of their common name.

Spider mites generally live on the undersides of leaves where they cause damage by puncturing the plant cells to feed. This produces leaf mottling and, in severe cases, leaf loss. Heavily infested plants are severely weakened and may even die.

There are about 1,200 species, although the most commonly known is the two-spotted spider mite or glasshouse red spider mite (*Tetranychus urticae*), which attacks a wide range of plant species. It prefers hot, dry conditions, so is worse on indoor or glasshouse plants or during warm summers, and under optimal conditions (24–27°C / 75–80°F) it can produce huge populations that soon disfigure the plant. Eggs hatch in as little as three days and mites become sexually mature in as little as five days. Each female can lay up to 20 eggs per day and can live for two to four weeks.

Eelworms

These microscopic worm-like animals, also known as nematodes, live in the soil. Many species feed on bacteria, fungi and other microorganisms. Some are parasitic predators of insects and are used as biological controls (see p. 213).

The major eelworm plant pests include chrysanthemum eelworm (*Aphelenchoides ritzemabosi*), onion eelworm (*Ditylenchus dipsaci*), potato cyst eelworm (*Heterodera rostochiensis* and *H. pallida*) and root knot

There are numerous species of slugs that live in the average garden. Most species feed on a wide range of plant material. Some are predators and eat other slugs and worms.

PESTS, DISEASES AND DISORDERS

Tanacetum coccineum, pyrethrum

eelworms (*Meloidogyne* spp.). These live and feed within the plant, whereas others remain in the soil and attack the plants' root hairs. Some species also transfer plant viruses while they feed. Symptoms of attack include stunted, distorted brown and dying foliage or swollen stems. Affected plants may be weak, lack vigour or even die.

Mammals and birds

Larger animals can be voracious and undiscriminating grazers of plants. In the wild these can range from the smallest mice to the largest elephants of Africa. Gardeners are extremely unlikely to come across anything nearly as large as an elephant, which can only be a blessing as they can be extremely destructive in their search for food, uprooting entire trees in order to get to their nutritious parts.

Rabbits feed on a very wide range of ornamental plants, fruit and vegetables and can easily cause sufficient damage to kill herbaceous plants, shrubs and young trees. They can even ringbark mature trees. New plantings and soft growth in spring are most susceptible, even on plants that are not generally eaten at other times.

Gardeners need to be particularly wary of rabbits, which can graze herbaceous plants down to ground level, and damage the foliage and soft shoots of woody plants up to a height of 50 cm (20 in). Where rabbits are a problem, chicken-wire fencing should be erected to a height of 1.4 m (5 ft) and tree guards put in place around the main trunks of susceptible woody plants. The bottom 30 cm (12 in) of wire fencing should be bent outwards at right angles and laid on the soil surface to prevent rabbits from burrowing under it.

Deer, especially muntjac (*Muntiacus reevesi*) and roe deer (*Capreolus capreolus*), can cause severe damage to a wide range of plants. Damage is similar to that produced by rabbits, but being larger animals the damage is usually more extensive. While deer will eat most plants, especially recently planted ones, there are some, such as tree peonies, that are usually left alone.

Squirrels, particularly the grey squirrel (*Sciurus carolinensis*), attack a wide range of ornamental plants, fruit and vegetables. Rather than eat the foliage, they tend to eat fruits, nuts and seeds, flower buds (especially from camellias and magnolias), vegetables like sweetcorn, and they dig up and eat bulbs and corms and – most seriously – strip the bark from trees.

Birds can be serious garden pests, while others, particularly species of tits (*Cyanistes* and *Parus* spp.) can be useful pest controllers, feasting on insect pests. The wood pigeon (*Columba palumbus*) is usually the worst bird pest on plants, eating the leaves of a wide range of plants, particularly brassicas, peas, cherries and lilac. They peck at the foliage and rip off large portions, often leaving just the leaf stalk and larger leaf veins. They may strip buds, leaves and fruit from blackcurrants and other fruit bushes. Bullfinches (*Pyrrhula pyrrhula*) feed on wildflower seeds for most of the year, but when food becomes more scarce in late winter, they begin to feast on unopened buds, particularly those of fruit trees.

James Sowerby

1757–1822

James Sowerby was an English naturalist, engraver, illustrator and art historian. Over the course of his lifetime, he singlehandedly illustrated and catalogued thousands of species of plants, animals, fungi and minerals of Great Britain and Australia.

Sowerby enjoyed a dual career, achieving distinction both as an outstanding artist and as a keen scientist. He bridged the gap between art and science, working closely with botanists to make his drawings as accurate and as scientific as possible. His main aim was always to bring the natural world to a wider audience of gardeners and nature lovers.

He was born in London and, having decided to become a painter of flowers, studied art at the Royal Academy. He had three sons – James De Carle Sowerby, George Brettingham Sowerby I and Charles Edward Sowerby, who all followed their father's work and became known as the Sowerby family of naturalists. His sons and grandsons continued and contributed to the enormous volumes of work he began and the Sowerby name remains firmly associated with illustrations of natural history.

Sowerby's first venture into botanical illustration was with William Curtis, Director of the Chelsea Physic Garden, London, and he illustrated both *Flora Londinensis* and *Curtis's Botanical Magazine* – the first botanical journal published in England. Sowerby created as well as engraved the illustrations, 70 of which were used in the first four volumes. This and an early commission from the botanist L'Héritier de Brutelle to provide the floral illustrations for his large work on the *Geranium* family, *Geranologia*, and two later works, helped lead to Sowerby's prominence in the field of botanical illustration.

A Specimen of the Botany of New Holland, written by James Edward Smith, was both illustrated and published by Sowerby, and was the first monograph on the flora of Australia. It was prefaced with the intention of meeting the general interest in, and propagation of, the flowering species of the new Antipodean colonies, but also contained a Latin botanical description of the plant samples. Sowerby's hand-coloured engravings, based upon original sketches and specimens brought to England, were both descriptive and striking in their beauty and accuracy. This use of vivid colour and accessible texts was the beginning of Sowerby's over-arching intention to reach as wide as audience as possible with works of natural history.

At the age of 33, Sowerby began the first of several huge projects – a 36-volume work on the botany of England, *English Botany or Coloured Figures of British Plants, with their Essential Characters, Synonyms and*

James Sowerby illustrated thousands of species of plants and fungi. He worked with scientists to produce careful and meticulously detailed work.

Agaricus lobatus,
tawny funnel

Careful illustrations of fungi, mushrooms and toadstools were just one of Sowerby's specialities.

Telopea speciosissima,
waratah, kiwi rose

This illustration by James Sowerby was used in *A Specimen of the Botany of New Holland*, the first book on the flora of Australia.

Places of Growth. This was published over the next 24 years, contained 2,592 hand-coloured engravings, including numerous plants receiving their first formal publication, and became known as *Sowerby's Botany*.

Unlike other flower painters of the time, whose work tended toward pleasing wealthy patrons, Sowerby worked directly with scientists. His careful and meticulous description of the subjects, drawing from specimens and research, was in direct contrast to the flower paintings of the Rococo period found illuminating the books of that time. The appealing hand-coloured engravings, usually sketched quickly in pencil, became highly valued by researchers into new fields of science.

His next project was on a similarly huge scale – *The Mineral Conchology of Great Britain*, which was a comprehensive catalogue of invertebrate fossils of England. This was published over 34 years, latterly by his sons James and George. He also developed a theory of colour and published two landmark illustrated works on mineralogy: *British Mineralogy* and its supplement, *Exotic Mineralogy*.

Having such a scientific disposition, Sowerby retained as many of the specimens used in his work as possible. A comprehensive collection of his work is maintained at the Natural History Museum in London. It includes his drawings and specimen collection for *English Botany*, approximately 5,000 items from his fossil collection and a large collection of his personal correspondence. Further work is held by the Linnean Society of London.

Fungi and fungal diseases

Although sometimes regarded as plants, fungi are classified into their own kingdom, and they include all the mushrooms, toadstools and moulds. Genetic studies actually show that fungi are more closely related to animals than to plants. Two major differences between plants and fungi is that fungi contain no chlorophyll and their cell walls contain chitin, not cellulose. The study of fungi is called mycology. Two major plant diseases once thought to be fungi – *Phytophthora* and *Pythium* (water moulds) – are now classified as belonging to the kingdom *Chromista*.

The kingdom *Fungi* encompasses an enormous diversity of organisms, and like plants there is a massive variety in their lifestyles, life cycles and morphologies, ranging from single-celled aquatic species to large mushrooms. Gardeners tend to notice them only when they are fruiting, producing their characteristic mushrooms or toadstools. At other times of the year they often exist below the soil or are hidden within plant tissues. Larger fungal species produce a mycelium, consisting of a mass of branching, fine, thread-like 'roots'. Large mycelial masses are often found in soils containing lots of organic matter.

Fungi derive their nutrients from organic matter, such as waste and detritus from plants, animals and other fungi. Those that feed on dead or decaying organic matter are known as saprophytes, and they have a fundamental role in nutrient recycling, particularly in the soil. Other fungi may be parasitic on plants (or animals); many of these are serious diseases of cereal crops, or have a less damaging symbiotic relationship, such as those that form mycorrhizal associations. The group of fungi known as truffles are mycorrhizal, associated with the roots of trees like beech, hazel, birch and hornbeam.

Pathogenic fungi

The fungi that cause diseases are called pathogenic. Some, such as grey mould (*Botrytis cinerea*), are ubiquitous and will attack a wide range of plants. Others, such as powdery mildews and rusts, have a narrow, specific host range, often comprising either one or a few related plant species. Although there are numerous fungal diseases, each one is specific to a species or group of species. The rust that attacks snapdragons is a different species from the one that attacks roses, and the powdery mildew affecting grapes is a different species from that attacking peas. Gardeners must still be careful, however, not to spread diseases around the garden on infected tools.

Fungal pathogens colonise living plant tissue and obtain their nutrients from the living host cells. Necrotrophic fungal pathogens infect and kill host tissue and extract nutrients from the dead host cells. Pathogenic fungi are often grouped together depending on the type of symptoms they cause, such as wilts, but there are often numerous genera and species of fungi that cause these symptoms.

Carpinus betulus, hornbeam

Common problem diseases caused by fungi

There are numerous fungal pathogens, some much more serious than others. Fairy rings in lawns (*Marasmius oreades*), for example, can be tolerated as cosmetic blemishes, whereas a bracket fungus on a large tree may indicate serious deterioration of its internal structure. A few fungi are found right through the plant, but most are restricted to localised plant tissues. Listed below are some of the commonest you will find in the garden.

Grey mould

Grey mould (*Botrytis cinerea*) is so called because of its greyish fuzzy growth. It is very common and can live on most living and dead plant material. It attacks the above-ground structures of plants, usually gaining entry through wounds or areas of damage. It also infects plants under stress, but it will infect healthy plants as well, especially under humid conditions. Other species of grey mould found in the garden include *B. galanthina* on snowdrops, *B. paeoniae*, which causes peony wilt or peony bud blast, *B. fabae*, which cause chocolate spot of broad beans and *B. tulipae*, which causes tulip fire.

Good hygiene is very important to keep grey mould at bay, particularly if plants are grown under glass. Remove any dead or dying plant material promptly, and make sure plants in the glasshouse are well ventilated and not overcrowded.

Rust diseases

Rust diseases (*Puccinia* spp. and other genera) are so called because of the usually rusty-brown colour of the spores and pustules, and they attack a wide range of garden plants. The spores vary in colour, according to the rust species and the type of spore that it is producing. For instance, rose rust produces orange pustules during summer, but in late summer and

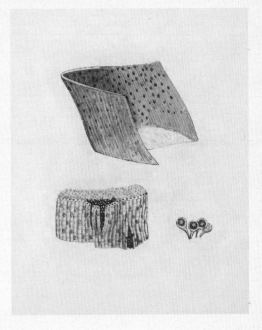

Leek rust (*Puccinia allii*) is a common leaf disease that also attacks onions, garlic and closely related plants. It can reduce yield and affect crops in store.

autumn these are replaced by black pustules containing overwintering spores.

Plants commonly affected include hollyhocks (*Alcea*), alliums, antirrhinums, chrysanthemums, fuchsias, mahonias, mints (*Mentha*), pelargoniums, pears (*Pyrus*), roses and periwinkles (*Vinca*). Rusts usually have two hosts, which they alternate with during their life cycle. European pear rust, for example, spends part of its life on junipers.

Rust diseases are unsightly and often, but not always, reduce plant vigour and in extreme cases can kill the plant. Leaves are most commonly affected, but rust can also be found on stems, flowers and fruit. Severely affected leaves often turn yellow and fall prematurely. Infection is favoured by prolonged leaf wetness, so rusts are usually worse in wet summers.

To deter rust, provide conditions that encourage strong growth, but avoid an excess of fertiliser as this can result in a lot of soft new growth that is easily colonised by rust. Remove all dead and diseased material at the end of the growing season to reduce the chance of spores overwintering. If rust is seen on leaves early in the growing season, its progress can be slowed by picking off affected leaves as soon as they are seen. Removing large numbers of leaves, however, is likely to do more harm than good.

Powdery and downy mildews

Powdery mildews are a group of related fungi (*Podosphaera* spp. and other genera) that attack a wide range of plants, causing a white, powdery coating on leaves, stems and flowers. Many plants are affected including apples, blackcurrants, grapes, gooseberries, peas and ornamentals such as asters, delphiniums, honeysuckles (*Lonicera*), oaks (*Quercus*), rhododendrons and roses. The plant tissue sometimes becomes stunted or distorted, such as leaves affected by rose powdery mildew.

Mulching and watering plants is an effective method of control as it reduces water stress and makes plants much less prone to infection. Removal of infected material promptly will also hinder reinfection.

Downy mildews spoil the appearance of ornamental plants and affect the yield and quality of edible crops. Unlike powdery mildew, downy mildew is not so easy to recognise. Typical symptoms include discoloured blotches on the upper leaf surface with a corresponding mould-like growth on the underside, with leaves shrivelling and turning brown, as well as stunted growth and lack of vigour. They are diseases of wet weather, as infection needs prolonged leaf wetness. A range of common edible and ornamental plants can be affected, including brassicas, carrots, grapes, lettuces, onions, parsnips, peas, foxgloves (*Digitalis*), geums, hebes, busy Lizzies (*Impatiens*), tobacco plants (*Nicotiana*) and poppies (*Papaver*).

Control downy mildew by the prompt removal and disposal of affected material (either bury or burn it, or put it in your council green waste bin) and avoid dense planting to ensure good air circulation. In glasshouses, open the windows to improve ventilation. Avoid watering plants in the evening as this can increase humidity around plants and, with it, the possibility of infection.

Blackspot

Blackspot is a serious disease of roses, caused by *Diplocarpon rosae*. It infects the leaves and greatly reduces plant vigour. Typically, a purplish or black patch appears on the upper leaf surface. The leaf may turn yellow around this spot and the leaf often drops; badly affected plants can shed almost all their leaves. The fungus is genetically very diverse and new strains arise very quickly. This means that disease resistance bred into new cultivars usually fails to last.

Collect and destroy, or bury, fallen leaves in the autumn to help delay the onset of the disease the following year. Use of fungicides can be more effective, and the advice is to alternate several products to maximise their effectiveness.

Delphinium, delphinium

Cankers and wilts

Cankers are generally round or oval areas of dead, sunken tissue, often starting at a wound or at a bud. Some cankers are caused by bacterial pathogens (see pp. 205–206). Apple canker is caused by *Neonectria galligena*, which attacks the bark of apples and some other trees. Willow black canker affects stems and is caused by *Glomerella miyabeana*. Parsnip canker is mainly caused by *Itersonilia perplexans*.

To help prevent cankers, pay attention to a plant's cultivation requirements, making sure that its drainage needs and soil pH requirements are met. Any affected areas should be completely cut out well below the point of infection and destroyed. Cankers on trunks and large branches are much harder to control; exposure of the canker to dry conditions by removal of the outer bark around the infected part may cause it to dry out and heal. Consider removing badly infected plants.

Wilting of plants, if not caused by drought, is often caused by fungal attacks. There are numerous different fungal species that can cause wilts, and in turn they attack a wide range of plants. Species include *Verticillium* and *Fusarium*, which attack herbaceous and woody plants. Those that are more specific to their host include *Phoma clematidina* – clematis wilt. *Verticillium* wilt shows itself by the yellowing and shrivelling of lower leaves, with some or all of the plant suddenly wilting, especially in hot weather. Brown or black streaks are seen in the tissue under the bark. Pay particular attention to weed control as some weeds are hosts for the disease, and be careful to not spread the fungus around the garden in contaminated soil. *Fusarium* wilt causes stunted growth and yellowing and wilting of the leaves on a number of ornamental plants. A reddish discolouration of the xylem vessels is an indicator of infection, as are white, pink or orange fungal growths and root and stem decay. Remove and dispose of infected plants promptly.

Clematis wilt causes rapid wilting of clematis and in severe cases can kill the entire plant. Resistant cultivars are available, but infection can be deterred by making sure the plants are given a good start on planting with a deeply cultivated hole and with a layer of mulch to reduce root stress. Remove and destroy all wilted growth back to healthy stems; new healthy shoots may subsequently form at ground level.

Leaf spots

Leaf spots are also commonly associated with a number of physiological disorders (see pp. 217–218), so it can be very difficult to tell whether the symptoms are caused by a fungus or some unconnected environmental factors. Common fungal leaf spots include hellebore leaf spot caused by *Microsphaeropsis hellebori*, primula leaf spot caused by several *Ramularia* fungi, viola leaf spots caused by *Ramularia lactea*, *R. agrestis* and *Mycocentrospora acerina* and currant and gooseberry leaf spots caused by *Drepanopeziza ribis*.

Strawberries are affected by various fungal diseases: *Mycosphaerella fragariae* (common leaf spot), *Diplocarpon earliana* (leaf scorch), *Phomopsis obscurans* (leaf blight) and *Gnomonia fruticola*.

Remove and destroy affected leaves. Infection can be deterred by ensuring that the cultivation requirements of plants are met to encourage strong growth. Affected currant and gooseberry bushes should be well fed and mulched to reduce water stress – some cultivars show levels of resistance.

Honey fungus

Honey fungus is the common name given to several different species of *Armillaria* fungus that attack and kill the roots of many woody and perennial plants. The most characteristic symptom is a white fungal sheath (mycelium) between the bark and wood, usually at ground level or just above, which smells strongly of mushrooms. Clumps of honey-coloured toadstools sometimes appear on infected stumps in autumn, and black 'bootlaces' (rhizomorphs) may be found in the soil. Honey fungus spreads underground, attacking and killing the roots of perennial plants and then decaying the dead wood. It is one of the most destructive fungal diseases in gardens.

To prevent honey fungus from spreading to unaffected plants, put in place a physical barrier made of a rubber pond liner 45 cm (18 in) deep into the soil and protruding 3 cm (1¼ in) above the soil. This, and regular deep cultivation, will break up and deter the spread of the rhizomorphs. If honey fungus is confirmed, remove and destroy, or send to landfill, the affected plant, including the roots, which will otherwise continue to feed the fungus.

Phytophthora

Phytophthora pathogens include some of the most destructive plant diseases, including *Phytophthora infestans*, which causes tomato and potato blight. Other *Phytophthora* species include *P. ramorum*, which has more than 100 host species, and *P. cactorum*, which causes rhododendron root rot and bleeding canker in hardwood trees.

Phytophthora infestans (potato blight) starts on the foliage, but transfers via the soil to the tubers, where it causes a destructive rot, making the potatoes inedible.

Phytophthora root rot is caused by several *Phytophthora* species and, after honey fungus, is the most common cause of root and stem decay in garden trees and shrubs. Herbaceous perennials, bedding plants and bulbs can also be affected. *Phytophthora* root rot is primarily a disease of heavy or waterlogged soils, and the symptoms can be very difficult to distinguish from those arising from just waterlogging – they include wilting, yellow or sparse foliage and branch dieback. *Pythium* species cause root rots in numerous plants and when they kill emerging seedlings it is known as 'damping off'. The fungus *Rhizoctonia solani* also causes damping off.

These are very difficult to control, but improving the soil drainage can greatly reduce the risk of plants succumbing to *Phytophthora*. Where the disease is new or localised in the garden, affected plants should be destroyed and the soil replaced with fresh topsoil. In the UK, where *P. ramorum* is suspected, it should be reported to the Plant Health and Seeds Inspectorate (PHSI).

Viral diseases

The existence of viruses was discovered by Dmitri Ivanovsky, a Russian biologist, in 1892. He was trying to find a bacterial cause of the tobacco mosaic disease, which was causing great problems at the time to tobacco crops. In the end he discovered a particle much smaller than a bacterium, which he called an 'invisible disease'. This was subsequently named a virus, and it was only with the invention of the electron microscope in the 1930s that viruses could actually be observed. They are between 20 and 300 nm (nanometers) in diameter.

Viruses cannot be classed as living organisms, because they do not grow (they only reproduce), they do not respire, and they are not cellular. Some scientists refer to them as 'mobile genetic elements', since they are simply a core of nucleic acid surrounded by a sheath of protein. On infection, viruses switch off the host cell DNA and use their own nucleic acids to instruct the cell apparatus to manufacture new viruses. Viruses, therefore, do not have the capability to replicate without the presence of a host, and so they are obligate parasites.

Although there are only a few viruses that are seen with any frequency in the garden, there are roughly 50 plant virus families, grouped into more than 70 genera. Viruses are named according to the symptom they cause in the first host in which they were found. For example, the first virus to be discovered also infects potatoes, tomatoes, peppers, cucumbers and a wide range of ornamental plants, but it is still known as *Tobacco Mosaic Virus* (TMV).

Although viruses are systemic – found throughout an infected plant – their symptoms may be seen only in one area or may vary from one part of the plant to another. Plant viruses affect many plants and cause distorted growth and discoloration of leaves, shoots, stems and flowers, as well as a loss of vigour and reduced yields, although they rarely kill the plant.

Transmission of viruses

To be successful, viruses have to be able to spread from host to host. Since plants do not move around, plant-to-plant transmission usually involves a vector. With TMV, man is the chief vector; the virus is very stable and heat-resistant, so it remains unaffected in tobacco and can be transmitted on the hands of smokers.

A similar mode of transmission occurs in commercial tomato houses, and in any propagation environment, when sap is directly transferred from an infected plant to a healthy one during pruning or propagation, via a knife or even on the fingers. A relatively small number of viruses can pass through infected seed.

Insects are the commonest vector, particularly sap-sucking insects such as aphids, leafhoppers, thrips and whiteflies. TMV has been observed to travel on the jaws of leaf-eating insects; it is an extremely infectious virus that only needs the breaking of a single leaf hair to cause eventual infection of the whole plant. Soil-inhabiting nematodes (eelworms) can transmit viruses as they feed on infected roots, and fungal pathogens can also transmit viruses.

Viruses vary in their ability to survive outside their host; many can

Cucumis sativus, cucumber

Common viral diseases

Virus name	Host	Vector
Cauliflower Mosaic Virus (CaMV)	Members of the *Brassicaceae* family; some strains infect members of the *Solanaceae*	Aphids
Cucumber Mosaic Virus (CMV)	Cucumbers and other cucurbits as well as celery, lettuce, spinach, daphnes, delphiniums, lilies, daffodils (*Narcissus*) and primulas	Aphids and infected seeds
Lettuce Mosaic Virus (LMV)	Spinach and peas, as well as ornamentals, especially osteospermums	Aphids and infected seeds
Tobacco Mosaic Virus (TMV)	Potatoes, tomatoes, peppers, cucumbers and a wide range of ornamental plants	Thrips; also mechanically on tools or fingers and occasionally through infected seed
Tomato Spotted Wilt Virus (TSWV)	Wide range of plants, apart from tomatoes, including *Begonia*, *Chrysanthemum*, *Cineraria*, *Cyclamen*, *Dahlia*, *Gloxinia*, *Impatiens* and *Pelargonium*.	Thrips, especially the western flower thrips.
Pepino Mosaic Virus (PepMV)	Tomatoes	Mechanically transmitted, although seed transmission is possible.
Canna Yellow Mottle Virus (CaYMV)	Cannas	Not known, but it is thought it may be transmitted mechanically, such as on propagating tools
Bean Yellow Mosaic Virus (BYMV)	Beans	Several aphid species

tolerate a wide range of temperatures, but they may not survive outside their host for long. Many are rapidly killed by exposure to heat and sunlight, whereas others are robust enough to be transmitted via pruning tools. A few can even survive composting.

There are no chemical controls for plant viruses, apart from trying to control or limit the insect vectors with an insecticide. Non-chemical control includes destroying infected plants promptly to prevent them becoming a source of further infection. Some ornamental plants and weed species are affected by the same viruses, so keep weed growth to a minimum. Wash and disinfect hands and tools that have come into contact with infected plants. Never propagate from plants suffering virus infection.

BOTANY IN ACTION

Colour 'breaking' in tulips

Viruses are a significant cause of poor growth, but sometimes the deleterious effect they have on plant health can have ornamental consequences. In tulips, streaking or 'breaking' in flowers can be caused by tulip breaking virus, a disease spread by aphids. Though there are mild and severe strains of the virus, all forms have a degenerative effect on bulbs. Today, the huge array of tulips available that bear variegated flower colour is the product of selective breeding and has a genetic, rather than pathological, basis.

Tulips showing various colour breaks.

Bacterial diseases

Bacteria are microscopic, single-celled organisms, which reproduce asexually by binary fission – each cell splitting into two. This process can occur as often as once every 20 minutes, so large colonies soon build up. Most bacteria are motile and have whip-like flagella that propel them through films of water. Along with fungi, they are the main decomposers of organic matter in the soil.

Approximately 170 species of bacteria can cause diseases in plants. They cannot penetrate directly into plant tissue, but enter through wounds or natural openings, such as stomata in leaves. Bacteria are tough, and if they do not find a host they can become dormant until an opportunity presents itself.

Ralstonia solanacearum is a bacterium that causes bacterial wilt or brown rot in potatoes. It was one of the first diseases shown to be caused by a bacterial pathogen.

In contrast to viruses that 'live' inside plant cells, bacteria grow in the spaces between cells, producing toxins, proteins or enzymes that damage or kill the plant cells. *Agrobacterium* causes cells to genetically modify, changing the levels of auxins produced, resulting in cancer-like growths, called galls. Other bacteria produce large polysaccharide molecules that block the xylem vessels, causing wilts.

The prevalence of bacteria

Bacteria are normally present on plant surfaces and will only cause problems when conditions are favourable for their growth and multiplication. These conditions include high humidity, crowding, and poor air circulation around plants.

Diseases caused by bacteria tend to be prevalent on plants during the winter months when light intensity and duration are reduced. During this time, plants are not growing actively and are easily stressed. Any condition that causes plant stress can make them more susceptible to infection, including fluctuating temperatures, poor soil drainage, a deficit or surfeit of nutrients, and incorrect watering. Misting plants will provide a film of water on the leaves where bacteria can multiply.

Bacterial diseases are typically spread in rain splash, in the wind or by animals. People can also spread them by using infected tools, by incorrect disposal of infected plant material or by managing plants poorly during winter. Symptoms of infection are usually localised but, as many cause rapid deterioration of plant tissues, their effects can be quite dramatic. They include tip burn, leaf spot, blight, canker, rot, wilt or the total collapse of plant tissues.

In some cases, the infection is evident as a bacterial ooze, such as in bacterial canker, clematis slime flux and fireblight. Many such infections, which result in the softening of plant tissues, are accompanied by a characteristic, and often unpleasant, smell. They also bear evocative names.

Bacterial canker

Bacterial canker is a disease of the stems and leaves of *Prunus* species, caused by *Pseudomonas syringae*. It causes sunken patches of dead bark, cankers, often accompanied by a gummy ooze, and small holes in the leaves, called 'shothole'. If the infection spreads all round the branch it will die.

Crown gall

Crown gall is a disease of the stems and roots of many woody and herbaceous plants. Infection causes knobbly swelling (galls), on the stems, branches and roots. It is caused by *Agrobacterium tumefaciens*.

Blackleg and bacterial soft rot

These are infections of potato tubers, caused by *Pectobacterium atrosepticum* and *P. carotovorum*. Tubers develop a soft and often foul-smelling rot. The blackleg bacterium also causes a soft rot at the base of the stem, leading to yellowing and wilting foliage.

Clematis slime flux

Clematis slime flux is caused by various species of bacteria, affecting most clematis species. It results in wilting, dieback and the appearance of a white, pink or orange, foul-smelling exudate from the stem. The disease can be fatal, but pruning out affected parts may occasionally save plants. Slime flux disease can also develop on the stems of a wide range of trees and shrubs, including *Cordyline*.

Fireblight

Fireblight is caused by the bacterium *Erwinia amylovora*. It only infects those members of *Rosaceae* in the sub-family *Maloideae*, such as apples, pears, cotoneasters, hawthorns (*Crataegus*), photinias, pyracanthas and *Sorbus*.

Symptoms include flowers wilting and dying at flowering time, with shoots shrivelling and dying as the infection spreads. Cankers are also seen on

Clavibacter michiganensis (syn. *Corynebacterium michiganense*) causes bacterial canker in tomatoes. The first symptoms are plants wilting followed by spots on leaves and fruit.

branches, and a slimy white liquid may exude from infections during wet weather. Severely affected trees may appear as though they have been scorched by fire. The disease was accidentally introduced to the British mainland from North America in 1957; it is now widespread, but in island areas such as the Isle of Man, the Channel Islands and Ireland, it is not yet established. Suspected outbreaks must be reported to the Plant Health and Seeds Inspectorate (PHSI).

Bacterial diseases are difficult to control. Emphasis instead should be on preventing the disease in the first place. Cultural control includes the use of sterile seed and propagation materials, disinfection of pruning tools and preventing surface wounds that act as portals to infection.

Parasitic plants

Some plants are unable to produce all or some of their nutrients themselves and are parasitic, or partially parasitic on other plants. More than 4,000 species of parasitic flowering plants are known to exist; they may be obligate – unable to complete their life cycle without a host – or facultative. Obligate parasites often have no chlorophyll at all, so lack the green colouring typical of most plants; facultative parasites do have chlorophyll and can complete their life cycle independent of the host.

Epifagus virginiana, beech drops

Striga coccinea, witchweed

Some parasites are generalists, such as dodder (*Cuscuta*) and red bartsia (*Odontites vernus*), being able to parasitise many plant species. Others are specialists that parasitise just a few or even one species, such as beech drops (*Epifagus virginiana*) on American beech (*Fagus grandifolia*).

Parasitic plants attach to either the stems or the roots of the host plant. They have modified roots, called haustoria (singular: haustorium), which enter the host plant and then penetrate the phloem, xylem or both to extract nutrients. Broomrape (*Orobanche*), dodder and witchweed (*Striga*) cause huge economic losses in a variety of crops. Mistletoes (*Viscum*) cause economic damage to forest and ornamental trees.

Broomrape

Broomrape (*Orobanche*) is a genus of more than 200 species of obligate parasites that completely lack chlorophyll and attach to the roots of their hosts. They have yellow or straw-coloured stems, leaves reduced to triangular scales and bear yellow, white or blue flowers that look like antirrhinums. Only the flowerheads are visible above the soil surface. The seedlings put out a root-like growth, which attaches to the roots of nearby hosts.

Dodder

Dodder (*Cuscuta*) is a genus of about 150 species of obligate parasites with yellow, orange or red stems and leaves reduced to minute scales. They have very low levels of chlorophyll but some species, such as *C. reflexa*, can photosynthesise slightly. Dodder seeds germinate at or near the surface of the soil and the seedling then has to find its host plant quickly. Seedlings use chemosensory methods to detect and grow towards their host (see chapter 8). The dodder seedling will die if it does not reach its host plant within 10 days.

Witchweed

Witchweed (*Striga*) are annuals that overwinter as seeds, easily spread by wind, water, and soil or animal vectors. Some species are serious pathogens of cereal crops and legumes, particularly in sub-Saharan Africa. In the USA, witchweed was considered such a serious pest that in the 1950s Congress allocated money in an attempt to eradicate it. This led to research that has since enabled American farmers to nearly eradicate it from their land. The seeds only germinate in the presence of exudates produced by the host's roots. After germination, they develop haustoria that penetrate the host's root cells, causing a bell-shaped swelling to form. Witchweeds colonise underground, where they may spend several weeks before emerging to flower and produce seeds.

Castilleja

Castilleja is a genus of about 200 species of annuals and perennials with brightly coloured flowers. The majority are natives of North America, and are commonly known as Indian paintbrush or prairie-fire. They are facultative parasites on the roots of grasses and other plants. Because of their highly attractive flowers, much research has gone into growing *Castilleja* in the garden or as greenhouse plants without a host plant. The best species to try are *C. applegatei*, *C. chromosa*, *C. miniata* and *C. pruinosa*.

Rafflesia

Rafflesia is an extraordinary genus of approximately 28 species found in southeastern Asia. They are obligate parasites of *Tetrastigma* vines. The only part of the plant that can be seen outside the host are the huge flowers, which in some species are over 1 m (3 ft) in diameter, weighing up to 10 kg (22 lb). Even the smallest species, *R. baletei*, has flowers that are 12.5 cm (5 in) in diameter. The flowers smell and even look like a dead, rotting animal, which gives rise to their common names: corpse flower and meat flower. The rotting smell attracts insects, especially flies, which pollinate the flowers.

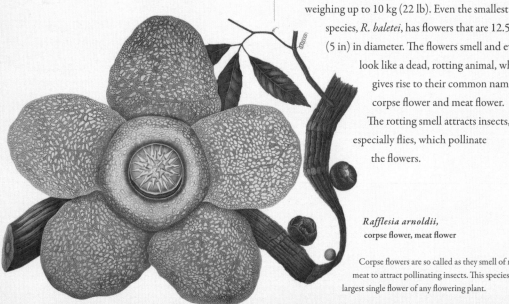

Rafflesia arnoldii,
corpse flower, meat flower

Corpse flowers are so called as they smell of rotting meat to attract pollinating insects. This species has the largest single flower of any flowering plant.

How plants defend themselves

Plant defences are either passive or active. Passive defences are always present in the plant, such as the stinging hairs of nettles (*Urtica dioica*), whereas active defences are only seen when the plant is injured, such as a chemical response. The one advantage of an active defence is that it is only produced when needed, and it is potentially less costly, therefore, with regard to energy spent by the plant on its production. As is to be expected, there is huge variation in defence mechanisms, which range from modified plant structures to the synthesis of toxins.

Mechanical defences

The first line of defence for any plants must be its surface layer.

Urtica dioica, nettle

Outer bark and cuticles
In woody plants, the surface defence would be its corky bark and lignified cell walls; for herbaceous plants it is likely to be the thick cuticles over their leaves and stems. They must be able to withstand a degree of physical attack.

Compound release to surface
Some defensive compounds are produced internally and released onto the plant's surface. For example, resins, lignins, silica and waxes cover the epidermis and alter the texture of the tissue. The leaves of *Ilex* (holly), for example, are very smooth and slippery, which makes feeding on them difficult; they are also quite hard and thick, and many are armed with sharply pointed teeth. Grasses have a high silica content, which can make them very sharp, unpalatable and indigestible; mammals known as ruminants (such as cattle and sheep) have evolved the ability to graze upon such plants.

Cutin
The cuticle contains an insoluble polymer called cutin, which is a very effective barrier against microbes. Some fungi, however, produce enzymes that are capable of breaking down cutin, and the cuticle can also be breached through openings such as the stomatal pores or through wounds, allowing microorganisms to enter.

External hairs, trichomes, spines and thorns

External hairs or prickles (called trichomes), spines or thorns are all present to prevent pests getting close to the plant. Trichomes can also possess insect-trapping barbs, or sticky secretions such as those seen on the leaves of the carnivorous sundew plant (*Drosera*), or they many contain irritants or poisons, such as those on cannabis plants that are rich in cannabinoids.

Raphides

Some cells contain raphides, needle-shaped crystals of calcium oxalate or calcium carbonate that make ingestion painful, damaging the herbivore's mouth and gullet and so ensuring more efficient delivery of the plant's chemical defences. Spinach plants are rich in calcium oxalate raphides. These would not be terribly good for you if consumed in large quantities; fortunately they are destroyed by cooking.

Chemical defences

Plants produce many different chemicals that help defend against attack. These include alkaloids, cyanogenic glycosides and glucosinolates, terpenoids and phenols. These substances are known as secondary metabolites as they are not involved in the primary functions of the plant: growth, development and reproduction.

Alkaloids

These chemicals include caffeine, morphine, nicotine, quinine, strychnine and cocaine. They all adversely affect the metabolic systems of animals that ingest them and produce a bitter taste, putting off the animal from eating them in the first place. The poisonous alkaloid taxine is found in all parts of the yew tree (*Taxus baccata*) apart from the red aril that surrounds the seed.

Cyanogenic glycosides

A considerable number are known to be toxic to one degree or another. Cyanogenic glycosides become toxic when herbivores eat the plant and break the cell membrane, releasing hydrogen cyanide.

Glucosinolates

Glucosinolates are activated in much the same way as cyanogenic glucosides and can cause severe abdominal problems and mouth irritation.

Terpenoids

The terpenoids include volatile essential oils such as citronella, limonene, menthol, camphor and pinene, plus latex and resins, which can be toxic to animals. These are responsible for making rhododendron leaves poisonous and include compounds such as digitalin, present in foxgloves (*Digitalis*).

Phenols

Phenols include tannins, lignin and cannabinoids. They make the plant difficult to digest, and interfere with the digestive processes. Pelargoniums produce an amino acid in their petals to defend against their main pest, Japanese beetles, which become paralysed after eating the petals. Toxalbumins are toxic plant proteins present in legume families and *Euphorbiaceae*.

Other secondary metabolites

Secondary metabolites do not just have poisoning roles. Flavonoids not only play important roles in the transport of auxin, root and shoot development and pollination, but they also possess antibacterial, antifungal and antiviral properties that help protect plants against infections.

Some of the secondary metabolites produced by plants are used as insecticides. These include nicotine from tobacco plants (*Nicotiana*), pyrethrin extracted from the flowers of certain chrysanthemum species, azadirachtin from the neem tree (*Azadirachta indica*), d-limonene from *Citrus* species, rotenone from species of *Derris* and capsaicin from chilli peppers.

Allelochemicals are secondary metabolites that are known to influence the development of neighbouring plants. The walnut (*Juglans nigra*) and the tree of heaven (*Ailanthus altissima*) are both known to secrete allelochemicals from their roots that may inhibit the growth of plants under their canopies. The leaf litter of some trees and shrubs is also known to have a similar effect. The mouse-ear hawkweed, discussed in chapter 4, is also known to be allelopathic.

Taste in plants

Many secondary metabolites have distinctive odours or tastes. These have no doubt evolved as a response to herbivory, but to humans they give many food plants their distinctive qualities. The chemical interactions involved are sometimes very complex, such as with tomatoes, which are often described as having anything from an earthy or musty flavour to a fruity or sweet tang. Furthermore, these flavours change as the fruit ripens, due to a complex mixture of volatile aromatic chemicals that also interact with fruit sugars, including fructose and glucose. Between 16 and 40 chemicals contribute to tomato flavours.

The fruit of the West African shrub *Synsepalum dulcificum* is known as the miracle fruit as it has the unique property of making sour food taste sweet. It contains a glycoprotein, called miraculin, that binds to the tongue's taste buds, making everything that is subsequently eaten taste sweet, even if it is sour. From the same part of the world comes the African serendipity berry (*Thaumatococcus daniellii*), which also produces an intensely sweet protein called thaumatin. It is 3,000 times sweeter than table sugar and contains almost no calories, making it useful as a natural sweetener for diabetics.

Many brassica crops, but particularly Brussels sprouts, have a bitter taste. In recent years many

Capsicum annuum,
chilli pepper, cayenne pepper

cultivars have been produced to reduce or remove this bitter taste and produce sweeter-tasting sprouts. The chemicals responsible for this bitter taste are groups of glucosinolates, used to deter insect and mammal herbivores. So the sweeter the Brussels sprout, the more susceptible it may be to pest problems.

Cucumbers were also once renowned for being quite bitter, and in some people they would cause quite severe indigestion and other problems with the alimentary canal. The chemicals known as cucurbitacins are responsible. While the 'burpless' cultivars of cucumber have been bred to produce fewer cucurbitacins, the production of cucurbitacins has been shown to increase under stress conditions. It is possible that these and other cultivars can become bitter when grown under unfavourable conditions, where there is insufficient or irregular watering, for example, or extremes of temperature and lack of nutrients.

Most legumes, but particularly soybeans, lentils, lima beans and kidney beans, are extremely toxic if eaten raw, because they contain chemicals called lectins. As a result, they have to be cooked, soaked, fermented or sprouted in order to make them safe to eat. Some legumes need to be eaten in relatively large amounts to cause toxic reactions, but just four to five raw kidney beans can cause severe stomach ache, diarrhoea and vomiting.

Further plant defences

Mimicry and camouflage

Plants employ many further tricks to deter their predators, which may lie outside the realm of active and passive defence. Mimicry and camouflage, for example, play a huge role. Thigmonastic movements, which occur in response to touch, are seen on the leaves of the sensitive plant (*Mimosa pudica*), which close up rapidly when touched or vibrated. The response is spread throughout the plant, and so the area of the plant exposed to attack is suddenly reduced. It may also physically dislodge small insects.

Some plants mimic the presence of butterfly eggs on their leaves, which has the effect of deterring female butterflies from laying real eggs. Some species of *Passiflora* produce physical structures resembling the yellow eggs of *Heliconius* butterflies on their leaves. Deadnettles (*Lamium*) mimic the appearance of real nettles (*Urtica dioica*) in the hope that predators will be fooled by this case of identity theft, and move on.

Good camouflage is less easy for a plant than it is for an animal, since plants often have to balance their need to 'hide' with their need to be attractive to pollinators and seed dispersers. The living stones of African deserts (*Lithops*) seem to achieve this very well, looking very much like the small stones and pebbles that it hides amongst.

Mutualism

Mutualism is another category of defence, where a plant attracts other animals to protect it from attack. Plants sometimes employ the services of another animal to give them a competitive edge, as already seen with the bull's horn acacia (see p. 191). This may simply be through the production of nectar and other sugary substances as a food source for the helper; some extrafloral nectaries (nectaries found outside the flower and not used in pollination) serve this purpose; they have been observed on some passion flowers (*Passiflora*) to attract ants, which in turn deter egg-laying butterflies.

Some plants even provide protection and housing for their 'helpers', as seen in some *Acacia* species, which develop enlarged thorns with a swollen base, forming a hollowing structure that houses ants. They also produce nectar in extrafloral nectaries on their leaves as food for the ants. Trees from the genus *Macaranga* have hollow stems, also used for ant housing, as well as providing their exclusive food source.

Mutualism also seems to occur on a chemical level. Fungal relationships may develop whereby the fungus produces toxins that are harmful to plant predators, a mechanism that has been observed in grasses such as *Festuca* and *Lolium*. This saves the plant from producing its own secondary metabolites.

Secondary metabolites are often produced in response to insect attack;

Mimosa pudica,
sensitive plant

as well as having a poisoning role, some act as a warning signal to other plants, and others can attract the natural predator of the organism attacking them.

Insect responses to secondary metabolites

Over time, and after repeated attempts to poison their assailants, it is only to be expected that all plant predators will begin to show some degree of adaptation or tolerance to secondary metabolites. Adaptations include the rapid metabolism of toxins or their immediate excretion. Many mammals have the ability to overcome mild toxification if their diet is varied, and some insects are able to feed on nicotine-producing plants by feeding exclusively on fluid in the phloem vessels, which contain no nicotine.

Remarkably, many insects are able to gather plant toxins into their own bodies and use them as a defensive mechanism against their own predators.

Sawflies feed on pine trees and store up the resins from the pine needles in their gut. These resins provide the sawflies with protection from birds, ants and spiders, as well as some parasites. Larvae of the cinnabar moth (*Tyria jacobaeae*) feed on ragwort (*Senecio jacobaea*) and absorb its poisonous alkaloids, making themselves unpalatable to predators. The adult moth's bright red colour serves as a warning.

Biological control

The natural enemies of plant pests are sometimes used or encouraged by gardeners as an alternative to chemical control. It is known as biological control. In the open garden, insects such as ladybirds, lacewings and hoverflies can eat large numbers of aphids and other soft-bodied insects, and they have a significant (if largely unnoticed) role to play in keeping pest populations within bounds.

Gardeners have begun to use with increasing success many natural predators much as predatory mites, parasitic wasps or pathogenic nematodes to control pest populations. As most of these biological controls need fairly constant conditions and have specific requirements in terms of warmth and humidity, they are mainly used in glasshouses and conservatories. The nematodes used to control slugs (*Phasmarhabditis hermaphrodita*) and vine weevil larvae (*Steinernema kraussei* and *Heterorhabditis megidis*) are active at soil temperatures above 5^0C (41^0F) and can also be used outside. Nematodes used to control leatherjackets, chafer grubs, sciarid fly and carrot, onion and cabbage root flies and caterpillars need warmer soil temperatures of $12°C$ ($55°F$) or more.

The caterpillars of *Tyria jacobaeae*, the cinnabar moth, feed on ragwort and absorb its poisonous alkaloids.

Breeding for resistance

Many plant breeders aim to breed resistance to pests and diseases into new cultivars to protect them against attack and, consequently, to help reduce the amount of pesticides used to control the problems. Owing to the time and expense involved, most of the research is focused on economically important food crops.

Breeding for plant pest and disease resistance generally involves finding suitable resistant genetic material in wild species or among existing cultivars, which is then incorporated into other cultivars. For example, in apples, research is carried out to develop resistance to fireblight (*Erwinia amylovora*), powdery mildew (*Podosphaera leucotricha*), scab (*Venturia inaequalis*) and woolly apple aphid (*Eriosoma lanigerum*). The main sources of resistant material used in breeding programmes comes from crab apple species, such as *Malus floribunda*, *M. pumila* and *M.* × *micromalus*.

The process of breeding for pest and disease resistance is not dissimilar to breeding for ornament (see p. 119). It includes the following steps:

Identification
Wild relatives and old cultivars are regularly used for breeding material as they often carry a useful resistance trait. As a result, much is done to preserve genetic material of species and older cultivars in gene or seed banks.

Crossing
A cultivar with desirable traits, such as taste and yield, is crossed with a plant with a source of resistance.

Growing on
The population of new plants produced from the cross are grown in a pest- or disease-conducive environment, usually in a glasshouse. This may require artificial inoculation and careful selection of the pathogen, as there can be significant variation in the effectiveness of resistance to different strains of the same pathogen species.

Selection
Resistant plants are selected. Because plant breeders are also trying to improve other plant traits, mainly related to yield and quality, careful selection is essential to ensure these other characteristics are not lost.

Many perennial crops, including most fruit and potatoes, are propagated by vegetative reproduction. In these, and other cases, disease resistance can be improved by more advanced methods, whereby genetic material from a resistant species or cultivar is introduced directly into plant cells. In some cases, genes from completely unrelated organisms are introduced. This realm of science is known as genetic modification, and although many people are concerned about the environmental implications of such work, there is no doubt that this is an area of plant breeding that is here to stay.

Resistance is termed durable if it continues to be effective over numerous years of widespread growing. Unfortunately, some resistance readily breaks down as pathogen populations evolve to overcome or escape the resistance, so further research and breeding is constantly needed.

Malus floribunda,
Japanese crab apple

PESTS, DISEASES AND DISORDERS

Physiological disorders

Many problems in plants have nothing to do with pests or disease. They may instead be caused by environmental or cultural conditions, such as poor or strong light, damage from weather, waterlogging or a lack of nutrients. These have a direct effect on the functioning of the plant's workings and systems.

The first way of determining if a physiological disorder is to blame is by checking that the plant has the right environmental and soil conditions, and by checking for recent extremes in weather patterns, such as heavy rain, dry spells, late or early frosts and strong winds. Soil analysis may also help. Further information on environmental factors and their effects on plants can be found in chapter 6.

Weather damage

Storm, snow and frost

Plants killed by cold and frost may start to leaf up in spring several months later, but suddenly die. The reason is that the leaves shoot as normal, but if the roots have died there is no water uptake to replace that lost by the leaves. Frost and cold are major causes of damage to tender plants, although even otherwise hardy plants can suffer if new growth is exposed to a hard frost, especially following a period of warm weather. Despite the example above, symptoms typically appear overnight, with wilting top growth, stem dieback and discoloured buds. Frosted blooms often abort and do not produce fruit.

In spring, frost and cold damage can be prevented by ensuring that tender plants are planted out only after the risk of frost has passed, and that they are properly hardened off and acclimatised to outdoor

Arisaema triphyllum f. *zebrinum*, jack-in-the-pulpit, Indian turnip

This interesting plant is hardy, but needs protection from late spring frosts, which can damage the flowers.

conditions. Susceptible plants can be protected with horticultural fleece if frost is forecast. Cold, drying winds can also severely inhibit spring growth even without an actual frost, so adequate shelter or erecting windbreaks is important.

Drought, heavy rains and waterlogging

Drought can cause plants to suffer from water stress and wilt. Once a plant has suffered from drought, it may not recover, depending on whether there has been severe damage done to the roots. Adequate watering is needed during prolonged hot, dry periods. Watering

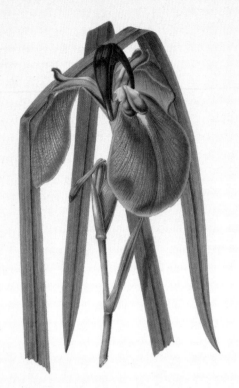

Iris ensata, the Japanese water iris, should be grown in moist to wet soil that does not dry out.

the soil around the roots, ensuring that the soil is thoroughly soaked a couple of times a week, is far more beneficial than shallow daily watering. Mulches also help to preserve soil moisture and they keep roots cool.

Heavy rains, particularly after prolonged dry periods, can cause root crops and tomatoes to split and potatoes to become deformed or hollow. Plenty of organic matter in the soil and using mulches will help to act as a buffer between sudden changes in conditions.

Waterlogging can occur on poorly drained clay soils, especially following heavy rainfall. Plants can become yellow and stunted, and they will tend to be more prone to any subsequent drought, as well as disease. Improving the soil and its drainage will help to alleviate this problem. Hail can cause damage to soft-skinned fruit, allowing brown rot and other diseases to infect the plant. Brown spot markings or lines on one side of an apple indicate a spring hailstorm.

A single, short weather event, such as a storm, snow or frost, can cause plant damage, but it is usually prolonged weather events that cause most damage, and in some cases, the symptoms might take weeks or months to show up. When faced with brown leaves, wilting, dieback or other symptoms, always consider the weather over the past twelve months, as well as looking for signs of pests or diseases.

Nutrient deficiencies

Poor growth and a number of problems, such as leaf discolouration, can be caused by a lack of plant nutrients in the soil. This may be due to shortages of the necessary nutrients, or even excesses, or because the nutrients are present but are 'locked up' and not available to the plant due to incorrect pH levels in the soil. The key to avoiding nutrient deficiencies is to ensure that the soil is healthy and contains plenty of well-rotted organic matter (see chapter 6).

While the main symptoms of nutrient deficiencies are explained on pp. 91–93, there are some that are specific to certain plants. Bitter pit in apples, for example, is caused by calcium deficiency; it causes the fruit skin to develop pits and brown spots, which taste bitter. Calcium deficiency in tomatoes and peppers also causes blossom end rot – sunken and dry, decaying areas at the blossom end of the fruit, furthest away from the stem.

Leaf spots

Although many fungal and bacterial diseases cause leaf spots, they can also be due to physiological disorders. This is particularly problematical on evergreens and plant species that are not completely hardy. The leaf spots usually show up as purple-brown spotting on the foliage and are typical of a plant under stress. Cold, wet winters, cold winds or frosty conditions, on their own or in combination, can all cause leaf spotting. Recently planted plants, especially mature or semi-mature specimens, are particularly susceptible.

Greenback

Hard green areas on tomato fruit are known as greenback, and blotchy ripening and internal areas of white or yellowish tissue are known as whitewall. Both are caused by excessive light, high temperatures or insufficient feeding. Widely fluctuating temperatures are likely to stress most plants, particularly cauliflowers, which will begin to show riciness, where individual florets develop and elongate, causing them to look like rains of rice.

Oedemas

Oedemas are raised corky spots and patches produced on the leaves. While they sound and look like a type of disease, they are actually caused by an excessive accumulation of water, when the plant takes up more water through its roots than it can readily lose through the leaves, leading to cell rupture. Overwatering or waterlogging are usually to blame, or growing plants in greenhouses or polytunnels where humidity is too high. It is most common in plants such as camellias, fuchsias, pelargoniums and cacti and succulents.

Crassula coccinea,
garden coral

As this is a succulent perennial, take care not to overwater, keeping compost on the dry side, especially when it is not flowering.

Mutations as disorders

Plant mutations, commonly known as sports, breaks or chimeras, are naturally occurring genetic mutations that can change the appearance of the organs of any plant. They can show up in a number of different ways: oddly coloured flowers, double flowers and foliage streaks and flecks or variegation.

Most mutations are random and are a result of a change within the cells, but they can be triggered by cold weather, temperature fluctuations or insect damage. More often than not, the plant reverts back to its original form the following year, but if the mutation is stable and carries on from year to year or from generation to generation, it has the potential to make a new, possibly commercially viable, cultivar.

Fasciation is a condition in flowering plants that produces flattened, often elongated, shoots and flowerheads that look like many stems compressed and fused together. It is brought about by abnormal activity in the plant's growing points, maybe as a result of random genetic mutation, infection by bacteria or viruses, damage caused by frost or animals, or even mechanical injury, such as hoeing or forking around the plant.

Occurrence of fasciation is unpredictable and is usually limited to a single stem. Plants commonly affected include delphiniums, foxgloves (*Digitalis*), euphorbias, forsythias, lilies, primulas, willows (*Salix*) and veronicastrums. Some stable fasciated plants are propagated to maintain their form and have become cultivated plants. These include *Celosia argentea* var. *cristata* and *Salix udensis* 'Sekka'. If you have fasciation in your garden, and it is not desired, it is simply removed using a clean pair of secateurs.

Vera Scarth-Johnson OAM
1912–1999

Vera Scarth-Johnson was a renowned botanist, botanical illustrator and conservationist. She is most fondly remembered for her love of the unique flora of the botanically rich area around Cooktown, and especially the Endeavour River Valley, on the Cape York Peninsula in Queensland, Australia. She was regarded as an Australian national treasure, an inspiration to many, and strove to inspire people to treasure the flora and habitats of this area of Australia.

Vera Scarth-Johnson was not only a much respected botanist and botanical illustrator, she was also an active conservationist.

Vera was born near Leeds in England, and went to school near the birthplace of Captain James Cook – the explorer who was the first recorded European to have visited Australia's eastern coastline. She went to finishing school in Paris, where she found little to interest her, except the garden there, and then studied art at two colleges of art in England.

Having been an avid gardener and botanist since childhood, she was keen to take up a career in horticulture. However, she could not find an employer willing to take on a female apprentice. Undeterred, after five years working at various jobs, she had saved up enough money to complete a horticultural course at the Hertfordshire Institute of Agriculture. After this she worked at a market garden until her grandfather, a well-to-do woollen manufacturer, gave her some money to start a market garden of her own.

In her mid-thirties, after World War II, Vera emigrated to Australia, possibly inspired by James Cook, living first in Victoria before settling in Queensland. Here she grew vegetables, tobacco and sugar cane, being only the second woman to obtain a sugar assignment. Vera was very much a hard-working, hands-on farmer.

During any spare time, Vera sketched and painted flowers of the local flora, amassing a great collection. In the mid-1960s, she heard a radio interview with the Director of the Royal Botanic Gardens, Kew, who mentioned how heavily the Gardens relied on voluntary assistance, especially from collectors around the world. Vera wrote to him offering her help, and enclosed some of her drawings, and so began her long and passionate association with the herbarium at Kew.

Vera made many journeys, all at personal expense, around Australia and the Pacific Islands collecting plant specimens for herbaria in Australia, the Royal Botanic Gardens, Kew and others in Europe and North America. They all benefited enormously from her research and passion for the subject, and the Queensland Herbarium received more than 1,700 plant specimens as a result.

Overcome with the beauty of the Endeavour River Valley, at the age of 60 Vera settled in Cooktown and began collecting the native plants of the region. Along with Aboriginals from the local Guugu-Yimithirr tribe, she made numerous extensive journeys discovering plant species and recording information on their uses. Possibly inspired by the botanical work of Joseph Banks, who took part in Captain James Cook's first great voyage of discovery, and Daniel Solander, the first university-educated scientist to set foot on Australian soil, she set out to paint and record the flora of the area. Unfortunately, the onset of Parkinson's disease meant she could only complete 160 illustrations.

Vera was an outgoing and charismatic character, and became an active and inspired campaigner against any developments that could adversely affect what she called 'my river'. Following a proposal to establish a silica sand mine on the north shore of the Endeavour River, the Endeavour River National Park was created after Vera alerted people to this threat.

Nicotiana tabacum,
tobacco

When Vera Scarth-Johnson moved to Queensland she started growing crops, including tobacco, as a source of income.

Vappodes phalaenopsis,
Cooktown orchid

The Cooktown orchid is the floral emblem of Queensland. This illustration by Scarth-Johnson shows it growing on a frangipani tree.

Her priceless collection of botanical illustrations is exhibited in the Nature's Powerhouse building in Cooktown Botanic Gardens. Her wish was that the collection and Nature's Powerhouse would encourage people to appreciate and protect the natural environment. The Vera Scarth-Johnson Wildflower Reserve is a nature area southeast of Bundaberg near the Kinkuna National Park.

National Treasures: Flowering Plants of Cooktown and Northern Australia is a collection of Vera's illustrations, her notes on them and a wealth of other information. Other books include *Wildflowers of the Warm East Coast* and *Wildflowers of New South Wales*.

Vera was awarded the Medal of the Order of Australia (OAM) for her contribution to art and the environment.

Bibliography

Attenborough, D. *The Private Life of Plants*. BBC Books, 1995.

Bagust, H. *The Gardener's Dictionary of Horticultural Terms*. Cassell, 1996.

Brady, N. & Weil, R. *The Nature and Properties of Soils*. Prentice Hall, 2007.

Brickell, C. (Editor). *International Code of Nomenclature for Cultivated Plants*. Leuven, 2009.

Buczaki, S. & Harris, K. *Pests, Diseases and Disorders of Garden Plants*. Harper Collins, 2005.

Cubey, J. (Editor-in-Chief). *RHS Plant Finder 2013*. Royal Horticultural Society, 2013.

Cutler, D.F., Botha, T. & Stevenson, D.W. *Plant Anatomy: An Applied Approach*. Wiley-Blackwell, 2008.

Halstead, A. & Greenwood, P. *RHS Pests & Diseases*. Dorling Kindersley, 2009.

Harris, J.G. & Harris, M.W. *Plant Identification Terminology: An Illustrated Glossary*. Spring Lake, 2001.

Harrison, L. *RHS Latin for Gardeners*. Mitchell Beazley, 2012.

Heywood, V.H. *Current Concepts in Plant Taxonomy*. Academic Press, 1984.

Hickey, M & King, C. *Common Families of Flowering Plants*. Cambridge University Press, 1997.

Hickey, M & King, C. *The Cambridge Illustrated Glossary of Botanical Terms*. Cambridge University Press, 2000.

Hodge, G. *RHS Propagation Techniques*. Mitchell Beazley, 2011.

Hodge, G. *RHS Pruning & Training*. Mitchell Beazley, 2013.

Huxley, A. (Editor-in-Chief). *The New RHS Dictionary of Gardening*. MacMillan, 1999.

Kratz, R.F. *Botany For Dummies*. John Wiley & Sons, 2011.

Leopold, A.C. & Kriedemann, P.E. *Plant Growth and Development*. McGraw-Hill, 1975.

Mauseth, J.D. *Botany: An Introduction to Plant Biology*. Jones and Bartlett, 2008.

Pollock, M. & Griffiths, M. *RHS Illustrated Dictionary of Gardening*. Dorling Kindersley, 2005.

Rice, G. *RHS Encyclopedia of Perennials*. Dorling Kindersley, 2006.

Sivarajan, V.V. *Introduction to the Principles of Plant Taxonomy*. Cambridge University Press, 1991.

Strasburger E. *Strasburger's Textbook of Botany*. Longman, 1976.

Websites

Arnold Arboretum, Harvard University
www.arboretum.harvard.edu

Australian National Botanic Gardens & Australian National Herbarium
www.anbg.gov.au

Backyard Gardener
www.backyardgardener.com

Botanical Society of America, web resources
www.botany.org/outreach/weblinks.php

Botany.com
www.botany.com

Chelsea Physic Garden, London
www.chelseaphysicgarden.co.uk

Dave's Garden
www.davesgarden.com

Garden Museum
www.gardenmuseum.org.uk

International Plant Names Index (IPNI)
www.ipni.org

New York Botanical Garden
www.nybg.org

Royal Botanic Gardens, Kew
www.kew.org

Royal Horticultural Society
www.rhs.org.uk

Smithsonian National Museum of Natural History, Department of Botany
www.botany.si.edu

University of Cambridge, Department of Plant Sciences
www.plantsci.cam.ac.uk

University of Oxford Botanic Garden
www.botanic-garden.ox.ac.uk

USDA Plants Database
www.plants.usda.gov

Index

abscission 161
air plants 50
alder-leaf birch 110
algae 12–13
aloes 5
Alpini, Prospero 60–61
Alyogyne hakeifolia 9
American bluebells 29, 133
angelica 72
angiosperms 25–28, 74
animals 110, 111, 147, 195
annuals 37
aphids 192
apical dominance 162
apomixis 115
apple trees 34, 173, 188, 214
arrowhead 68
asexual reproduction 102
Asteridae 26
Australian oak 53
Aztec lily 183

bacterial diseases 205–206
banana 60, 99
baneberry 40
Bauer, Ferdinand Lukas 116–117
Bauer, Franz Andreas 116
beardtongue 172
bee orchid 127
bee pollination 111, 113
beech drops 207
beech 70
bell peppers 78
biennials 37
binomial names 9, 30
biological controls 213
birds 195
birthroot 145
blackleg 206
blackspot 200
blue plaintain lilies 143
blueberries 144

bog bilberries 144
botanical names 29–30
bracken 20
brazil nut trees 80
breeding of plants 118–121
broomrape 207
bryophytes 14–15, 77
buds 51–53, 162–163
bugbane 40
bulbs 50, 82
Burbank, Luther 108–109
burning bush 185
buttercups 83, 111
butterfly bush 170

camellias 59, 91
Camerarius, Rudolf Jakob 101
camouflage 212
cankers 201, 206
Castilleja 208
castor-oil plant 124
catechu tree 126
cedars 24
cells and cell division 86–88
chemical defences 210–212
cherry trees 65, 163
Chinese swamp cypress 23
Chinese wisteria 165
cinnabar moth 213
clematis slime flux 206
climate 153–157
club mosses 19, 21
coconut 75, 79
codling moths 190
coffee plant 61
colour, response to 180
columbine 29, 133
common names 29
compatibility in pollination 114
conifers 22, 23, 24
Convention on International

Trade in Endangered Species (CITES) 24
Cooktown orchids 219
coppicing 174
cordyline 103
corms 50, 82
corn 33, 57, 118, 121, 186
cotyledons 124
crocuses 28
cross-pollination 112–113
crown gall 206
cryptophytes 50–51

cucumber 203
Culpeper, Nicholas 9
cultivars 39
cultivated oat 98
cultivation 118–121
Culver's root 157
Curtis, John 151
cut healing 165
cuttings 105–107
cycads 7, 22, 24
cypress trees 23, 24

daffodils 57, 176
dahlias 4
daisy, common 71
dandelion 58, 74, 115
Darwin, Charles 9, 168, 178, 181
dawn redwood 23
deadheading 167
deciduous 23, 70
defence mechanisms 209–213
delphiniums 200
diatoms 12
dicotyledons 28, 64, 75, 124
Dillhoffia cachensis 25
Dioscorides, Pedanius 8
diploids 15
disbudding 166

disease resistance 214
division 103
dodder 184, 208
dog lichens 18
dog violets 112
dogwood, red-barked 174
dormancy, seeds 125, 126
downy mildews 200
dragon arum 112
drought tolerance 149
Echinocactus 149
eelworms 194–195
elephant hawk moth 191
embryos 115, 128
epiphytes 50
ethnobotany 134
evergreens 23, 70, 171

F1 hybrids 40, 120–121
false cypress 24
false larch 23
family names 30
feeding 152, 167
ferns and relatives 19–21, 74
fertility, soil 145
field horsetail 21
filbert, large 175
fireblight 206
firs 24, 52
flame azaleas 155
flowering plants 25–28
flowers 71–73, 172
food storage organs 82–83
formative pruning 171
Fortune, Robert 54–55
foxgloves
    71, 111, 113, 127, 128
French lavender 102
fruit 78–81, 92, 93, 172–173
fruit trees, yields 144
Fuchs, Leonard 30
function 43
fungi 197, 198–202
gametophytes 14, 19–20

garden coral 217
genus 30, 34–35
geophytes 50
geotropism 181
germination 126–129
girdling 65, 164
golden chain tree 138
grafting 104–105
granny's bonnet/nightcap 29, 133
grapevines 119
grasses 27
greater snowdrop 107
greenwood cuttings 107
grex names (gx) 40
grey mould 199
ground ivy 90
group names 40
growing conditions, influencing 157
growth 43, 44–50
growth rings 65
guelder roses 193
gymnosperms 22–24, 74

haploids 15
hardiness 154
hardwood cuttings 105, 107
hellebores 133
herbaceous plants 26
Himalayan balsam 80
Himalayan screw pine 131
history 7, 8–9
holly 27, 113, 161
honey fungus 202
honeysuckle 8, 185
hornbeam 198
horsetails 19, 21, 104
hortensia 106
human senses 187
humus 145
hybridisation 36, 120–121
hybrids 39, 40
hydrangea 106
hydrophytes 50
Indian turnip 215–216

insects 190–193, 213
International Code of Nomenclature
    for algae, fungi and plants (ICN)
    29, 31
International Code of
    Nomenclature for Cultivated
    Plants (ICNCP) 29
ivy 57

jack-in-the-pulpit 215–216
Japanese iris 100, 216
Japanese knotweed 104
Japanese sago palm 24
junipers 23, 24
Kirk's tree daisy 40
kiwi rose 197
knotted kelp/knotted wrack 12
Krauss's spike moss 21

laburnum 138
larch 23
leaves 66–70, 107, 147, 201
Leek rust 198
Leyland cypress 23
lichens 18
light 128, 157, 178–181
lilac 122
lime 146–147
Lindley, John 150–151
Linnaeus, Carl 9, 30
liverworts 14–15, 77
lobster claw 141
London Plane 36
loose silky bent 156
lungwort 180

macronutrients 91–93
magnolias 26, 64
maidenhair tree 22, 24, 34
maize 33, 57, 118, 121, 186
mammals 195
maples 63–64, 79, 91, 161
Martens club moss 19
Masson's heath 117

INDEX

McClintock, Barbara 32–33
mechanical defences 209–210
meiosis 32, 88, 134, 135
Mendel, Gregor Johann
    16–17, 120
messmate 53
microclimate 153–157
micronutrients 93
mimicry 212
mint 63
mistletoe 35
mitochondria 90
mitosis 88, 124
moisture 148–149
monkey puzzle trees 24, 38
monocotyledons 26, 28, 64, 124
Monterey cypress 23
morning glory 179
mosses 14–15, 77
mutations 121, 217
mutualism 212–213

natural selection 118
nectar 190
nettles 209
nitrogen 92, 146
Nootka cypress 23
North, Marianne 168–170
nutrients 96–97, 152, 216–217

oak trees 44, 65, 70, 132, 175
offsets 103
organic matter 145, 147
oriental plane 36
ornamental cabbage 147
ovules, development of 115
oxygen 127

parasitic plants 207
passionflowers 35, 181
pathogenic fungi 198
peach trees 6
peas 17, 99
perennials 37

pests 190–195, 214
pH values, soil 144
phloem tissue 97
photomorphogenesis 178, 180
photoperiodism 179
photosynthesis 89–90
phototropism 178
physiological disorders 215–217
phytophthora 202
pinching out 166
pineapple 81
pines trees 24
pitcher plant 35
Plant Breeders' Rights (PBR) 41
Plant Health and Seeds Inspectorate
    (PHSI) 206
plant hormones 98–99
plant nutrition 91–93
plant selection 118–119
plum trees 42, 109, 166
plum yews 23
pollarding 174
pollen 190
pomegranate 75
pond algae 13
poppies 27
potatoes 142, 205
powdery mildews 200
pruning 159–167, 170–175
pteridophytes 19–21
pyrethrum 26, 195

queen sago palm 24
quince 31
quinine 76
quinoa 134

Rafflesia 208
rainwater 148–149
raspberries 78
red-barked dogwood 174
red clover 69
Redouté, Pierre-Joseph 182–183
reproduction 101–121

respiration 90
resurrection plant 21
rhizomes 50, 82, 83, 102
rhododendron
    55, 95, 130, 131, 144
ringing 65
roots 56–59, 106, 107, 166
rosebay willowherb 191
rosemary 171
roses 31, 53, 103, 120, 151, 158, 183
royal ferns 20
runners 83, 102, 104
rust diseases 199

sage-leaved rock rose 160
Sargent, Charles Sprague 94–95
saturated soils 148
saving seed 134–135
scarification 126
Scarth-Johnson, Vera 218–219
scent 184–185
secondary hosts, pests 193
secondary metabolites 210–212, 213
seed banks 74, 134–135
seed producers 22, 25
seeds 74–75, 78–79, 124–135, 190
self-incompatibility in
    pollination 114
self-pollination 112–113
self-seedlings, trueness of 133
semi-ripe cuttings 107
senescence 161
sensitive plant 212
sexual reproduction 110–115
Siberian lily 10
silky-leafed osier 67
silky wind grass 156
Silver Cycas 7
Silvery crane's-bill 154
sitka spruce 95
slugs 194
Smith, Matilda 130–131
snails 194
sneezewood 167

223

soaking seeds 127
soft rot 206
softwood cuttings 106
soil 138–149, 143
sound 186
Sowerby, James 196–197
sowing seeds 132–133
Spanish turpeth root 117
species 30, 36–37
spermatophytes 74
spider mites 194
sporangia 19
spores 19–20
sporophytes 14
sports 121
Spruce, Richard 76–77
spruce trees 24
St James lily 183
staggered germination 126
steinless gentian 86
stems 62–65, 105, 174
stolons 83
Strangman, Elizabeth 120
stratification 126
strawberries 27, 104, 201
Striga coccinea (witchweed) 207, 208

subspecies (subsp. or ssp.) 38
swamp cypress 23
Swedish whitebeam 115
sweet gum 63
sweet peas 125
sweet peppers 78

tawny funnel 197
taxonomy 11, 29, 30
tea tree 55
temperature 127, 154
Theophrastus 8, 30
thigmotropism 180–181
tobacco plants 219
tomatoes 121, 152, 206
touch sensitivity 180
Tournefort, Joseph Pitton de 133
trade designations 41
transpiration 96–97
tree collars 164
tree ferns 20–21
trumpet vine/trumpet
    honeysuckle 153
tubers 50, 82, 83
tulips 37, 50, 136
varieties (var) 38

vegetables 92
vegetative reproduction 102–107
Venus flytrap 34, 66
vibrations 186–187
viper's bugloss 72
viral diseases 203–204
waling-waling 151
waratah 197
water 96–97, 127, 148
water ferns 20, 21
water shoots 173
weather 153–157, 215–216
Weigela 140
wheat 121, 124
wild arum 67
wilting point 148–149
wilts 201
wind 110–111, 181
witch hazel 41

xerophytic plants 149
xylem vessels 96–97

yew trees 23, 39
yolk-yellow prosthechea 73

Image credits

Front cover: Huffcap Pear © RHS, Lindley Library
Digitalis purpurea © RHS, Lindley Library

Back cover: Rosaceae, Pyrus aria © RHS, Lindley Library

26, 29, 47, 48, 59, 60, 61, 65, 66, 67, 71, 72, 73, 75, 76, 78, 80, 86, 90, 91, 92, 99, 102, 103, 104, 119, 120, 125, 127, 128, 130, 142, 149, 151, 152, 153, 160, 163, 173, 179, 180, 181, 183, 193, 197, 200, 204 & 219 © RHS, Lindley Library

94, 168 & 182 © Alamy

96 © Getty Images

201 205 & 206 images used with permission of the Agricultural Scientific Collections Trust, NSW, Australia

All images in this book are public domain unless otherwise stated.

Every effort has been made to credit the copyright holders of the images used in this book. We apologise for any unintentional omissions or errors and will insert the appropriate acknowledgment to any companies or individuals in subsequent editions of the work.